Internal Wave Hydrodynamics on Submersibles

Peng Du · Chao Wang · Sen Zhao · Jun Wen ·
Luo Xie · Haibao Hu

Internal Wave Hydrodynamics on Submersibles

Volume 1: Interactions Between Internal Waves and Structures

Springer

Peng Du
School of Marine Science and Technology
Northwestern Polytechnical University
Xi'an, China

Sen Zhao
School of Marine Science and Technology
Northwestern Polytechnical University
Xi'an, China

Luo Xie
School of Marine Science and Technology
Northwestern Polytechnical University
Xi'an, China

Chao Wang
Marine Design and Research Institute
of China
Shanghai, China

Jun Wen
School of Marine Science and Technology
Northwestern Polytechnical University
Xi'an, China

Haibao Hu
School of Marine Science and Technology
Northwestern Polytechnical University
Xi'an, China

ISBN 978-981-95-4478-3 ISBN 978-981-95-4479-0 (eBook)
https://doi.org/10.1007/978-981-95-4479-0

This Springer imprint is published by the registered company Springer Nature Singapore Pte Ltd.
The registered company address is: 152 Beach Road, #21-01/04 Gateway East, Singapore 189721, Singapore

Preface

Internal solitary waves are widely-present in the density-stratified ocean and are a common type of nonlinear, large-amplitude wave phenomenon. During their propagation, they induce shear flow in the water above and below the wave surface, as well as strong convergence and divergence of water and sudden strong currents. Submersibles are highly likely to encounter internal solitary waves during underwater operations. The high crests and deep troughs of internal solitary waves, along with their concentrated energy, can severely affect the motion stability and hydrodynamic load characteristics of submersibles, potentially leading to instability and loss of control. This book focuses on the interaction between oceanic internal solitary waves and submersibles, an engineering application background. A three-dimensional stratified flow numerical wave tank is established using numerical methods. Based on the theories of internal solitary waves under different stratification models and combined with the RANS turbulence model, numerical simulations of density-strongly-stratified and continuously-stratified internal solitary waves are conducted. The nonlinear wave-body coupling process between internal solitary waves and vehicles under real-scale and various working conditions is further simulated. The motion response and load characteristics of vehicles under the action of strongly-stratified and continuously-stratified internal solitary waves are analyzed. The effects of submersion depth, wave amplitude, and vehicle speed on the multi-degree-of-freedom motion and load characteristics of vehicles are summarized. The differences in the motion and load changes of vehicles under the action of strongly-stratified and continuously-stratified internal solitary waves are compared. The effects of internal waves on the efficiency of submersible appendages and the influence of parameters such as internal wave amplitude, appendage submersion depth, and appendage working conditions are explored. Based on this, self-propelled experiments are conducted in a laboratory environment by changing appendage parameters to investigate the motion patterns of suspended and self-propelled vehicles under the action of internal waves, providing references for control measures when vehicles encounter internal waves in engineering practice. The specific research content and chapter arrangement are as follows:

- Chapter 1 introduces the background and significance of the research in this work, and systematically summarizes the current research status of internal solitary wave characteristics, including theoretical studies, observational experiments, and numerical simulations. It also reviews the research on the interaction between internal solitary waves and fixed structures and submersibles.
- Chapter 2 discusses three types of internal solitary wave theories—KdV, MCC, and DJL—and their applicable ranges. It compares the waveforms, wave speeds, and characteristic wavelengths of these theoretical equations under current fluid parameters. The waveform of the MCC equation matches well with the density mid-value waveform of the DJL equation, with its characteristic wavelength lying between those of the KdV and mKdV theoretical solutions. These conclusions lay the foundation for the application of the internal wave theory in the subsequent chapters.
- Chapter 3 introduces the theoretical basis of computational fluid dynamics involved in this simulation. It details the underwater motion equations and loads on vehicles. The MCC and DJL equations are used as the numerical wave generation theories for strongly-stratified and continuously-stratified internal solitary waves, respectively, to establish a three-dimensional numerical wave tank. The wave generation time step independence is verified. It is found that, compared with strongly-stratified internal solitary waves, the horizontal velocity of continuously-stratified internal solitary waves changes gradually at the density jump layer, with a symmetrical flow field and smooth stream-lines, and no influence of tail waves. The basic dimensions of the vehicle and grid division are introduced. The grid independence and reliability of the overlapping grid are verified through three sets of structured grids. The vehicle grid is optimized based on the non-dimensional wall distance. Finally, the accuracy of the internal solitary wave load and fluid-structure coupling method is verified through a case of interaction between internal solitary waves and a suspended ball, laying the foundation for the interaction between internal solitary waves and vehicles in the following chapters.
- Chapter 4 conducts simulations of the interaction between strongly-stratified internal solitary waves and suspended and moving vehicles under various working conditions. It studies the effects of three key parameters—submersion depth, wave amplitude, and vehicle speed—on the motion response and load characteristics of vehicles. The effects of internal solitary waves on vehicles are obtained by analyzing the heave, sway, and pitch of vehicles, combined with changes in horizontal force, vertical force, and pitching moment. It is found that under the action of strongly-stratified internal solitary waves, vehicles located above and at the interface are more significantly affected. The closer the vehicle is to the interface, the greater the impact on the load characteristics. Wave amplitude mainly affects the speed of the vehicle's motion response and the peak value of the load. The larger the wave amplitude of the internal solitary wave, the faster the vehicle's motion response and the greater the peak load. Additionally, vehicle speed has a significant impact on the peak value of the load and the duration of wave-body interaction.

- Chapter 5 conducts simulations of the interaction between continuously-stratified internal solitary waves and suspended and moving vehicles under various working conditions. It studies the effects of submersion depth, wave amplitude, and vehicle speed on the motion response and load characteristics of vehicles in a real marine environment. Based on this, the differences in the coupling interaction between strongly-stratified and continuously-stratified internal solitary waves and vehicles are compared. It is found that under the action of continuously-stratified internal solitary waves, the effects of submersion depth, wave amplitude, and vehicle speed on the vehicle's motion response and load characteristics are similar to those of strongly-stratified internal solitary waves. The main difference is that the motion and load of vehicles under the action of continuously-stratified internal solitary waves show stronger oscillation, and the load peak is smaller than that under the action of strongly-stratified internal solitary waves.
- Chapter 6 introduces the propeller model, computational domain, and grid division used in this work. The grid independence of the simulation is verified through three sets of grids with different scales. The numerical simulation of the open-water experiment of the propeller is conducted to verify the reliability of the numerical model and simulation method. Based on this, the density of the flow field is continuously stratified using a temperature-stratified flow model to study the hydrodynamic characteristics of the propeller under stratified conditions, which is found that the density stratification has a minor impact on propeller performance, with the propeller's thrust coefficient, torque coefficient, and open-water efficiency values being slightly higher than those in uniform flow only when the advance coefficient $J \geq 1.0$. Subsequently, the interaction between internal solitary waves and the propeller is numerically simulated using the established internal solitary wave numerical wave tank. Internal solitary waves have a significant impact on the performance parameters and wake field of propellers, especially when the propeller crosses the front wave surface and is located above the trough. The thrust and torque can drop sharply by 22.42% compared to the initial thrust, posing a significant threat to navigation safety.
- Chapter 7 introduces the rudder model, computational domain, and grid division used in this work. The accuracy of the numerical method is verified by simulating the non-dimensional lift and drag coefficients of the rudder structure under different rudder angles. Subsequently, the interaction between internal solitary waves and the rudder structure is numerically simulated using the internal solitary wave numerical wave tank to study the effects of key factors such as submersion depth, rudder angle, and wave amplitude on the efficiency of the rudder structure. It is found that rudder structures located below the wave surface experience severe oscillations in force curves and much larger force amplitudes than other positions, and are prone to flow separation, posing a significant threat to navigation safety. In internal wave environments, the stall angle of rudders at various depths is significantly advanced. When encountering internal solitary waves, the rudder angle should be less than 15°, preferably around 10°. The wave amplitude has little effect on the variation pattern of the internal solitary wave force on the rudder structure.

- Chapter 8 introduces the internal solitary wave experimental wave tank used in this work. Based on the experimental wave tank, a self-propelled experimental design is carried out. According to the size of the wave tank, the scale of internal waves, and the designed working conditions, the functions that the vehicle model needs to achieve are determined, and a vehicle model that meets the size and functional requirements is designed and manufactured. On this basis, a series of experiments on the interaction between internal solitary waves and suspended and self-propelled vehicles are carried out to explore the effects of key parameters such as submersion depth, vehicle speed, wave amplitude, and vehicle attitude on the motion response of vehicles. It is found that for suspended vehicles, those located below the trough are more likely to experience a "falling deep". For moving vehicles in the upper layer, an increase in wave amplitude reduces the drop depth but makes the pitch angle change more violently. Vehicles with a bow-down attitude have the largest range of vertical velocity changes and the most unstable motion. When located at the density interface, the larger the wave amplitude, the greater the drop depth. Vehicles at the interface are more sensitive to pitch angle changes. Vehicles with a bow-down attitude experience a rapid drop after passing through the trough, with the pitch angle reaching twice the initial value during motion.
- Chapter 9 briefly summarizes the research results of this work, points out the existing problems, and proposes future work prospects, providing a valuable reference for the simulation research on the interaction between internal solitary waves and submersibles.

This book simulates the interaction between strongly-stratified and continuously-stratified internal solitary waves and freely moving vehicles under various working conditions using numerical methods. It analyzes the mechanism of strong nonlinear wave-body coupling, summarizes the effects of submersion depth, wave amplitude, and vehicle speed on vehicle motion and load, explores the im-pact of internal solitary waves on the performance of vehicle appendages, and conducts research on the interaction between internal solitary waves and self-propelled vehicles. It provides a reference for control measures when vehicles encounter internal solitary waves in engineering practice and has certain engineering application value.

Xi'an, China Peng Du
Shanghai, China Chao Wang
Xi'an, China Sen Zhao
Xi'an, China Jun Wen
Xi'an, China Luo Xie
Xi'an, China Haibao Hu

Contents

Chapter 1
Introduction

1.1 Background

The 21st century is the century of the ocean. The ever—moving ocean, which covers about 71% of the earth's surface, is rich in mineral, oil and gas, energy, aquatic products and space resources. With the rapid increase of the world population and the gradual increase of energy consumption, countries around the world have turned their development focus to the ocean [1]. China is a major maritime country with 3 million square kilometers of jurisdictional sea areas and 18,000 km of coastline. Building a strong maritime country cannot be separated from the innovation and progress of marine science and technology equipment. From the "Jiaolong" to the "Fendouzhe", China has successively broken through a number of deep—diving technologies, finally turning the dream of "diving into the ocean to catch turtles" into reality; unmanned deep—diving series of underwater robots and underwater gliders such as "Hailong", "Qianlong" and "Haiyi" have also made great contributions in the fields of marine environmental observation and scientific investigation of deep—sea resource exploration and development (as shown in Fig. 1.1); Submersibles have a large self—sufficiency and endurance, and can operate independently in a large marine area for a long time away from the base [2].

The vast ocean holds infinite opportunities, but wave loads also pose a serious threat to common marine structures and submersibles [3–4]. The marine environment is complex and changeable. Although the surface often appears calm, there are hidden currents in the ocean's interior [5]. Internal waves are a common type of wave phenomenon. When disturbed by external forces and complex topography, they can undergo nonlinear polarization at the density interface and evolve into internal solitary waves. With the development of satellite remote sensing and in—situ observation technologies, the hazards of internal solitary waves have gradually become well—known [6–8]. Internal solitary waves are a special type of nonlinear oceanic internal wave that exists widely around the world [9], especially in the South China Sea, where internal wave activity is frequent [10–12].

P. Du et al., *Internal Wave Hydrodynamics on Submersibles*,
https://doi.org/10.1007/978-981-95-4479-0_1

Fig. 1.1 New—type marine equipment

(a) Jiaolong

(b) Haiyi

Internal solitary waves have large amplitudes and carry huge amounts of energy. They can maintain a stable waveform and propagate steadily. During propagation, they cause the seawater above and below the density interface to flow in a shearing state, triggering strong convergence and divergence of seawater and sudden strong currents (wave—induced currents) [13–17]. The amplitude of internal solitary waves ranges from several meters to hundreds of meters. The largest amplitude of internal solitary waves observed in the South China Sea so far is over 200 m. Their wavelengths extend from hundreds of meters to several kilometers, and the maximum flow speed of the induced flow field can exceed 2 m/s [18–20], as shown in Fig. 1.2.

When a submersible hovers underwater or travels freely, it will inevitably encounter different types of oceanic internal solitary waves. Due to the characteristic of internal solitary waves having high peaks and deep troughs, the vertical forces generated are extremely large. Two internal wave shear flows with opposite directions are formed in the water bodies of the upper and lower density layers. Their destructive power is sufficient to tear apart any objects they affect, thus posing a significant threat

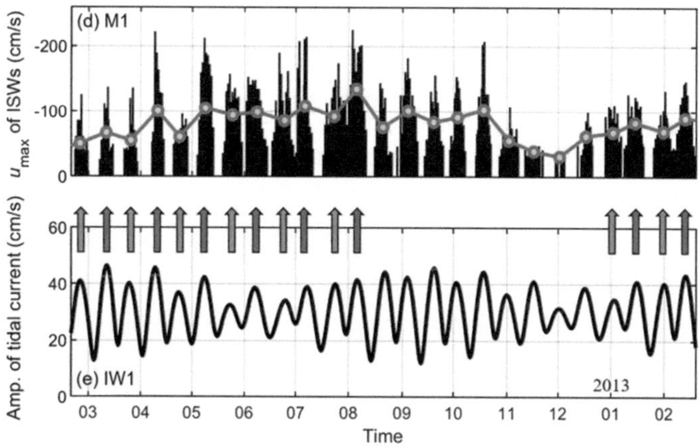

Fig. 1.2 The maximum westward flow velocity at the M1 mooring location in the South China Sea and the semi-diurnal tidal current amplitude variation at IW1 near Badan Island [19]

to marine engineering structures, especially submersibles that cross the density interface [21]. In engineering practice, accidents caused by internal solitary waves are not uncommon. There have been several news reports of submersibles being wrecked by internal solitary waves. On April 20, 2021 (20 April UTC), the Indonesian Navy submarine (KRI Nanggla—402) sank near the Lombok Strait, 100 km north of Bali (see the magenta star in Fig. 1.3). NASA proved that a powerful "underwater wave" occurred in this dangerous area, hitting the submarine and causing it to sink [22].

Internal solitary waves in the ocean have become a disastrous environmental factor that must be considered for the safe underwater navigation of submersibles.

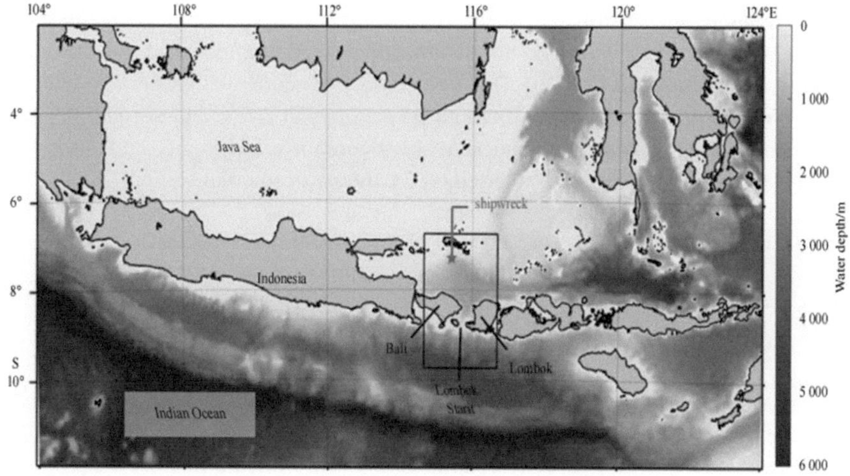

Fig. 1.3 Wreckage of the submarine found in the Lombok strait [22]

This book conducts research on the strong nonlinear interaction between internal solitary waves and submersibles. A three—dimensional numerical wave flume is established through numerical simulation methods to carry out research on the interaction between internal solitary waves and submersibles under full—scale and multiple working conditions. The motion response and load characteristics of submersibles under the action of internal solitary waves are analyzed to reveal their hazardous mechanisms and interaction laws. Further investigation into the impact of internal solitary waves on submersibles in a continuously stratified ocean environment can effectively enhance the safety and operational efficiency of submersibles. Meanwhile, as the propulsion and attitude control devices of the vehicle, the performance of the propeller and rudder also has an important impact on the navigation safety of the vehicle. Therefore, the influence of internal waves on the appendages of the vehicle should not be ignored and has strong engineering value, which can provide references for the maneuvering strategies of the vehicle when encountering internal waves.

1.2 State of the Art

In 1834, Scottish engineer Russell first observed solitary waves while working and described them in his later research report as follows: "This is a round, smooth, and well-defined water wave that travels several miles along the river channel with a constant shape and speed, and finally disappears in the meandering river channel [23–24]." Subsequ"nt research has found that it is difficult to establish an exact mathematical definition for solitary waves, but they can be associated with dynamic models (nonlinear equations) and have the following characteristics: (i) The shape of the wave does not change during propagation; (ii) The wave speed of the solitary wave decays or approaches a constant at infinity; (iii) After strong interaction with other solitary waves, solitary waves do not exhibit superposition effects and will maintain their own characteristics [25–26]. Due to its unique fundamental properties, internal solitary waves have always been favored by scholars at home and abroad. A large number of studies in ocean observation, wave generation experiments, and numerical simulations have been carried out based on the theory of internal solitary waves, and great progress has been made.

1.2.1 Research on Internal Solitary Wave Theory

The theory of internal solitary waves has always been a hot topic widely concerned in applied mathematics, physical oceanography, and ocean engineering. In 1895, Korteweg and G. de Vries [27] derived the famous KdV equation under the assumptions of small amplitude and weak nonlinearity. This equation was the first to be used to describe the surface of internal solitary waves and has been widely applied in fluid dynamics, aerodynamics, and continuum mechanics. It describes how waves evolve under the combined influence of weak nonlinearity and weak dispersion and is one of the most classical theories of internal solitary waves. Subsequently, many scholars have studied the KdV equation for the propagation of internal solitary waves under different water depths based on the relative size relationship between water depth and internal solitary wave wavelength in the actual ocean. In 1966, Benney [28] and Benjamin [29] derived the governing equation for finite-amplitude internal solitary waves based on the Boussinesq equation, which is based on the assumptions of long waves and finite wave amplitudes, namely the shallow-water internal solitary wave theory. The shallow-water theory is applicable when the wavelength is much greater than the water depth. However, when the water depth is much greater than the wavelength or the lower layer water depth tends to infinite water depth, the shallow-water theory is not applicable. Therefore, Benjamin [30] and Ono [31] proposed the deep-water internal solitary wave theory applicable to the case where the lower layer water depth tends to infinite water depth based on the shallow-water theory.

Since the KdV equation is applicable to internal solitary waves with small amplitude and weak nonlinearity, when the ratio of wave amplitude to total water depth is greater than 0.1, the KdV theoretical solution deviates significantly from the measured values of internal solitary waves. To make the KdV equation more widely applicable, Funakoshi and Oikawa [32] added a cubic nonlinear term to the KdV equation to obtain the eKdV model. Michallet and Barthelemy [33] proposed the mKdV model, which overcame the shortcoming of the KdV equation being inapplicable when the ratio of upper and lower layer fluid thicknesses is close to the critical depth ratio, and obtained a model applicable to cases where the thicknesses of the upper and lower layer fluids are close. The eKdV and mKdV theories are still based on weak nonlinearity and weak dispersion and are applicable to weak and moderately nonlinear internal solitary waves. Choi and Camassa [34] established a two-layer Green-Naghdi model with strong nonlinearity and weak dispersion. Starting from the fully nonlinear Euler equations, they accurately approximated the dispersive terms to the first order under the conditions of flat bottom with finite and infinite water depths. Under the steady-state condition, this model is the same as the internal solitary wave theory established by Miyata using the conservation law, so it is called the MCC (Miyata-Choi-Camassa) theory. According to the existing literature, the waveforms, wave speeds, and flow fields obtained by the MCC theory are very close to the experimental values [35–36], as shown in Fig. 1.4.

When considering the stratified structure of the ocean, to simplify the model, the effect of the density jump layer is usually ignored, and it is regarded as consisting of two layers of uniform fluid with different densities that do not mix with each other. The two layers of fluid show a strong density jump at the interface, which is the commonly used two-layer model [37]. The KdV equation and the MCC equation are both theories derived under the assumption of the two-layer model, but the actual ocean environment has a density jump layer thickness, and the two-layer model cannot fully characterize the true properties of internal solitary waves. In recent years, the theory of internal solitary waves has gradually developed and improved from the two-layer model to the continuously stratified direction. Fructus

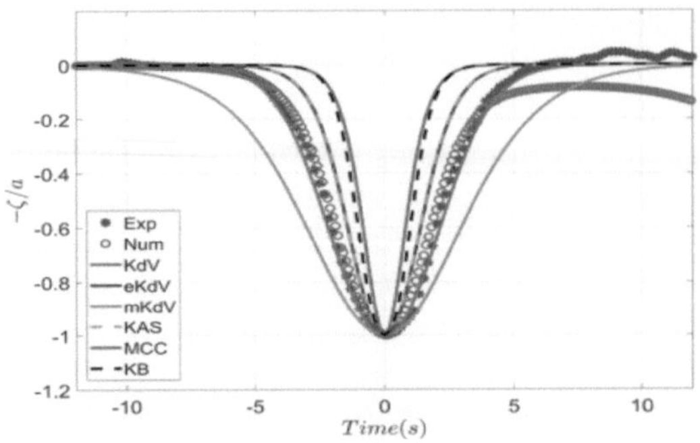

(a) Comparison of waveforms between numerical solutions, theoretical solutions, and experimental results [35]

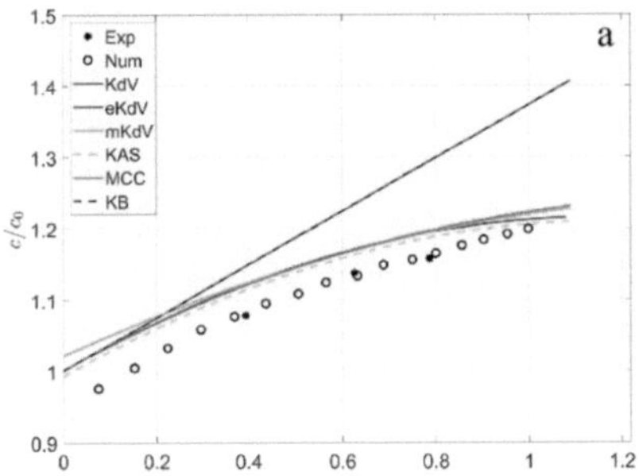

(b) Comparison of wave speeds between experimental results and theoretical models [35]

Fig. 1.4 Comparison of different internal solitary wave theories

and Grue [38] have derived a method for calculating fully nonlinear solitary waves propagating in a continuously stratified fluid, with the stratification determined by a constant Brunt-Väisälä frequency. To accurately describe the characteristics of large-amplitude oceanic internal solitary waves compared to the total water depth, Stastna [39] and Lamb [40] proposed the DJL (Dubreil-Jacotin-Long) equation, which is equivalent to the complete set of stratified Euler equations in a steady state without making assumptions about the nonlinearity of fluid flow. Therefore, the solutions obtained are exact internal solitary wave solutions in the inviscid limit and are applicable to fully nonlinear large-amplitude internal solitary waves [41]. Wang [42] abstracts the ocean as a system composed of three fluids with a stable density stratification. Based on irrotational flow, new strongly nonlinear and weakly nonlinear models are established, and the original equations are solved using MITgcm to confirm the existence and stability of the second-mode internal waves in the deep sea.

1.2.2 Observation and Experimental Research on Internal Solitary Waves

Due to the influence of solar thermal radiation, seawater forms a stable density stratification structure under the dominance of heat exchange. When internal solitary waves pass through, they can affect this stable stratification [43]. In the 1960s, with the development of thermistor vertical arrays (chains), the observation of large internal waves near the coast gradually became popular [44]. However, most of these observations were based on measuring the temperature changes in the water body after the passage of internal waves. With the rapid development of measurement technology and monitoring instruments, such as CTD (Conductivity Temperature Depth), current meters, acoustic Doppler current profilers, and various types of buoys and moorings, the three-dimensional observation of internal waves has been greatly enriched. During the Asian Ocean International Acoustic Experiment (ASIAEX), a large amount of field observation data on internal solitary waves in the northeastern South China Sea was obtained [45]. Klymak et al. [46] observed internal solitary waves with an amplitude of 170 m in the South China Sea using a fast CTD profiler. Xu et al. [47] observed an internal solitary wave in the South China Sea with a maximum amplitude of 207 m and a speed exceeding 2.9 m/s using moored ADCPs. Cai, through the analysis of a large amount of field observation data in the waters near the South China Sea, concluded the following [48]: First, in terms of wave modes, the internal solitary waves observed in the northeastern part of the Dongsha Islands can be divided into two modes, namely first-mode and second-mode internal solitary waves [49], with first-mode internal solitary waves being more active in summer than in winter; second-mode internal solitary waves in winter may be related to the thermocline being located at intermediate depths, and second-mode internal solitary waves are more active than first-mode internal waves. Second, the internal solitary wave packets observed from the Luzon Strait to the northeastern part of the Dongsha Islands can also be divided into two types [50]: In type-a wave packets, the largest internal solitary wave always leads other smaller internal solitary waves, while type-b wave packets usually consist of a single largest internal solitary wave at the center of the packet. Third, the internal wave energy characteristics in the northeastern South China Sea (four areas from the Luzon Strait to the continental shelf) are geographically different [51]. Synthetic Aperture Radar (SAR) remote sensing technology, with its characteristics of all-weather, all-time, high-resolution, and wide coverage, has been widely used in the observation of internal solitary waves [52], as shown in Fig. 1.5a. Due to the interaction between the surface waves caused by internal waves and ocean surface tension waves and short gravity waves, wave wakes are formed [53], which generally appear as alternating bright and dark stripes in SAR images [54]. These bright and dark stripes can provide basic parameter information of internal waves, such as wavelength, amplitude, wave speed, and propagation direction. Apel et al. [55] showed evidence of large internal wave packets appearing off the coast of New York Bay and the west coast of Africa, and proved that internal wave packets propagating towards the coast are separated by tidal periods, which is a common feature of coastal oceans, using SEASAT and coastal SAR images. Fu and Holt [56] used SAR images to show the surface characteristics of at least eight wave packets generated on the steep terrain between the Gulf of California and nearby islands, with internal waves appearing as alternating bright and dark bands, as shown in Fig. 1.5b.

Compared with field and satellite remote sensing observations, laboratory experiments on internal solitary waves are easier to operate, have lower costs, are highly repeatable, and allow for more intuitive observation of internal wave-related characteristic parameters. Scholars have conducted extensive research on laboratory wave generation methods for internal solitary waves, their propagation characteristics, and interactions with topography. Walker [57] established a stratified water tank and generated internal solitary waves by disturbing the interface between two layers of fluid with a wave paddle, further studying the propagation of internal solitary waves. Huang [58] used a dual-paddle wave generation method to conduct a series of experiments on the interaction between cylindrical structures and internal solitary waves, establishing a theoretical prediction model for internal solitary wave

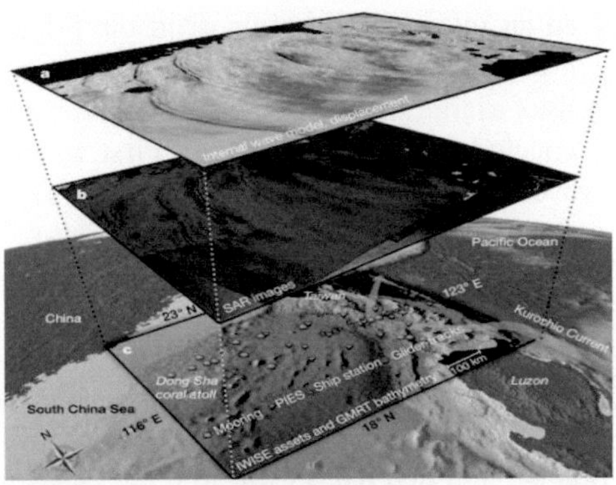

(a) Overview of internal waves in the South China Sea [52]

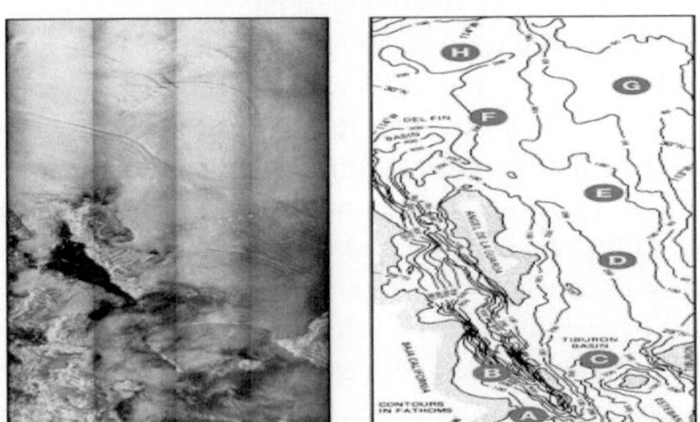

(b) Synthetic Aperture Radar (SAR) image of the northern Gulf of California [56]

Fig. 1.5 Internal solitary waves detected by synthetic aperture radar (SAR)

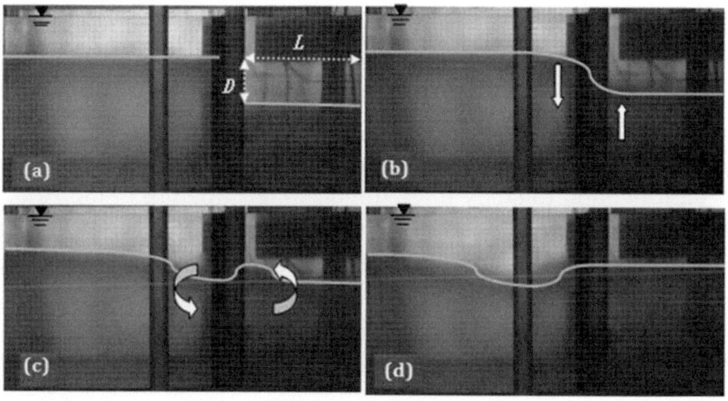

Fig. 1.6 Wave generation process by gravity collapse [64]

loads. Chen [59] et al. used the gravity collapse method in laboratory experiments to qualitatively describe the generation of internal solitary waves. Wei [60] et al. improved the traditional gravity collapse wave generation device and applied it to engineering. However, the detailed evolution process of internal solitary waves generated by gravity collapse has not yet been widely understood [61]. Chen [62] used experimental methods to study the behavior of internal solitary waves over irregular seafloor obstacles, with results indicating that the height of the obstacle significantly affects the degree of wave distortion. Tian [63] used a water tank experiment to study the penetration depth of internal solitary waves in partially saturated porous sandy silt and clayey silt seabeds, with experimental results showing that the instantaneous excess pore water pressure in the seabed decreases with increasing seabed depth. Du [64] studied the detailed process of generating internal solitary waves by gravity collapse (as shown in Fig. 1.6) and discussed the conditions for generating stable internal solitary waves in a laboratory water tank.

1.2.3 Numerical Simulation Research on Internal Solitary Waves

Computational Fluid Dynamics (CFD) is an emerging interdisciplinary field that combines fluid mechanics and computer science. With the rapid development of computers after the 1990s, CFD has grown swiftly and has become an important tool for fluid mechanics research, alongside experimental fluid mechanics. Compared to traditional laboratory wave tank methods, numerical simulations of internal solitary waves are more convenient and efficient in studying their generation, propagation, evolution, and dissipation. Moreover, solving the fluid control equations makes it easier to obtain other parameters in the flow field, such as pressure, velocity, and turbulent kinetic energy.

Grimshaw et al. [65], based on observed bathymetry and continuous stratification results in the South China Sea, used the Ostrovsky equation to study the impact of rotation on the evolution of shoaling solitary waves and conducted fully nonlinear simulations of shoaling internal solitary waves using MITgcm. Vlasenko and Stashchuk [66] used a fully nonlinear nonhydrostatic numerical model with continuous stratification to study three-dimensional shoaling in the Andaman Sea. Lamb [67] simulated the evolution of shoaling internal solitary waves in the South China Sea using a two-dimensional fully nonhydrostatic internal gravity wave model [68–69] and investigated the sensitivity of the predicted shoaling internal solitary wave motion to numerical resolution. Hu [70], based on nonlinear potential flow theory, studied the propagation and evolution characteristics of internal solitary waves over a sloping topography in a stratified fluid, established a numerical model for internal solitary wave propagation based on a multi-domain boundary element method, and validated it. Carr [71] used a numerical method combining contour advection and pseudo-spectral techniques to simulate shear-induced instabilities in internal solitary waves. Li [72–73] developed a numerical model—ISWFoam, using the OpenFOAM open-source software, to simulate internal solitary waves in continuously stratified, incompressible viscous fluids. ISWFoam accurately simulated the generation, evolution, breaking, and interaction of internal solitary waves with complex topography. Iwata [74] used a two-dimensional numerical simulation method to study energy dissipation caused by the breaking of internal solitary waves due to changes in the thickness of the density jump layer. Deepwell [75] used DNS to simulate that when internal solitary waves in continuously stratified fluids pass through a narrow side-wall contraction, vortices are generated above and below the density jump layer, as shown in Fig. 1.7a. Wang [76] used large eddy simulation to study the interaction between internal solitary waves and stepped topography, investigating the propagation, transmission, reflection, and destruction of internal solitary waves and analyzing the relationships between their geometric features, energy loss characteristics, and initial setup parameters. Hartharn-Evans [77] used the pseudo-spectral method SPINS numerical model to study the effect of stratification form on the shoaling characteristics of internal solitary waves propagating over a smooth linear slope under the same initial conditions. It was found that the stratification form affects the type of breaking of shoaling internal solitary waves, as shown in Fig. 1.7b.

1.2.4 Research on Interaction Between Internal Solitary Waves and Structures

Extensive theoretical, experimental, and numerical studies have shown that internal solitary waves are common oceanic wave phenomena, affecting the transport of mass and momentum in the ocean as well as energy transfer across different scales. Unlike other nonlinear internal waves, internal solitary waves have a single amplitude, maintain their waveform during propagation, and concentrate most of their energy

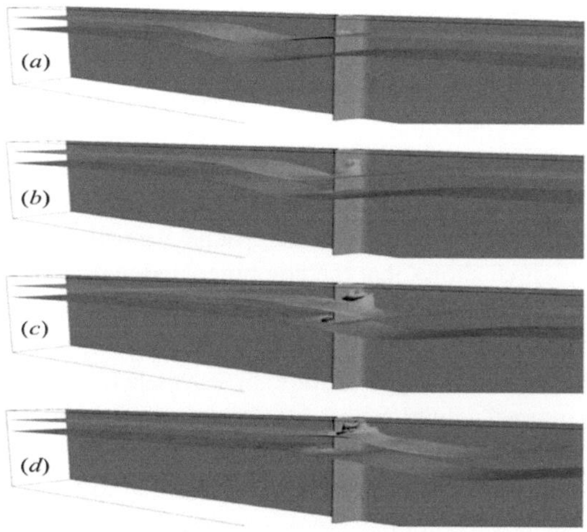

(I) Isosurface plot of large and small vortices formed above and below the density jump layer when internal solitary waves pass through a contraction [75]

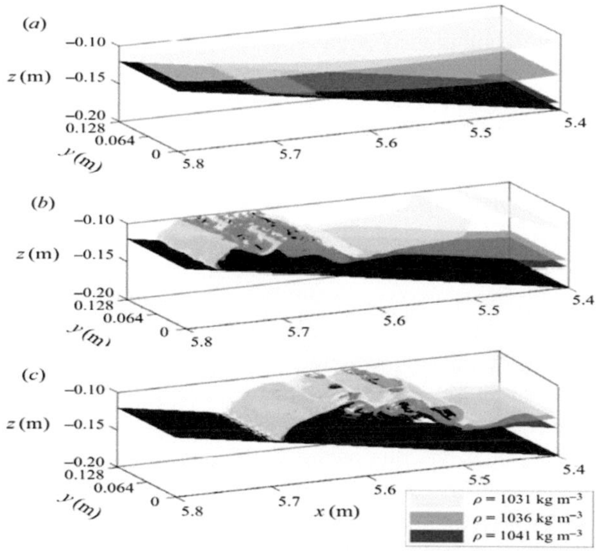

(II) Three-dimensional view of density isosurfaces showing blob-like vortices formed during the surging of internal solitary waves [77]

Fig. 1.7 Numerical simulation of internal solitary waves interacting with topography near the wave crest, posing a significant threat to marine structures [78]. Over the past few decades, scholars have conducted a large amount of research on the interaction between internal solitary waves and marine structures.

Cai et al. [79–80] first introduced the Morison [81] formula to calculate the load on a cylinder due to internal solitary waves and proposed that only the first mode needs to be considered in the calculation [82]. Subsequently, they studied the forces on a cylinder under the combined action of background shear flow and internal solitary waves [83]. The research shows that the combined action of background shear flow and internal solitary waves significantly increases the load on the cylinder. However, these studies are based on the assumption that nonlinear internal solitary waves can be approximated by a series of linear internal waves, which may be applicable to weakly nonlinear internal solitary waves but not to strongly nonlinear cases. Xie [84] used a model based on the fully nonlinear Euler equations in two dimensions to study the time-varying load on a cylinder in a continuously stratified ocean and used this model to investigate the effect of barotropic currents on internal solitary wave loads.

In experimental research, You's [85–87] research group conducted a series of experimental studies on the interaction between internal solitary waves and marine structures. They proposed combining the Morison formula with the Froude-Krylov formula to calculate full-scale structures and established predictive models for the loads on semi-submersible platforms, FPSOs, etc., due to internal solitary waves. These models were continuously improved through numerical validation. Zhang [88] experimentally studied the forces on a vertical cylinder due to internal solitary waves by measuring the forces on small segments of the cylinder at different depths to determine the vertical distribution of the internal solitary wave forces.

In numerical simulations, Song [89] used a time-domain numerical model for the interaction between internal solitary waves and marine structure motion responses to compare the excitation forces and responses of internal solitary waves and surface waves on Spar platforms. The results showed that the maximum excitation response caused by internal solitary waves is about three times that of surface waves. Wang [90] et al. used the commercial software Fluent to establish a numerical simulation method for the strong nonlinear interaction between internal solitary waves and semi-submersible platforms in finite water depth with a two-layer fluid. They verified that the internal solitary wave load on the semi-submersible platform consists of wave pressure difference force, viscous pressure difference force, and frictional force. Wang [91–92] et al. studied the load characteristics of circular and square columns under the action of internal solitary waves through model experiments and numerical simulations. They summarized the interaction laws of internal solitary waves with different amplitudes on columns. The results showed that as the amplitude of the internal solitary wave increases, the horizontal force on the column increases. When the amplitude is the same, the horizontal force on the square column is greater than that on the circular column. Cheng et al. [93] studied the propagation characteristics of internal solitary waves passing over marine risers and found that the amplitude of the transmitted wave and the lateral force on the riser increase with the increase of the internal wave amplitude and riser diameter, as shown in Fig. 1.8.

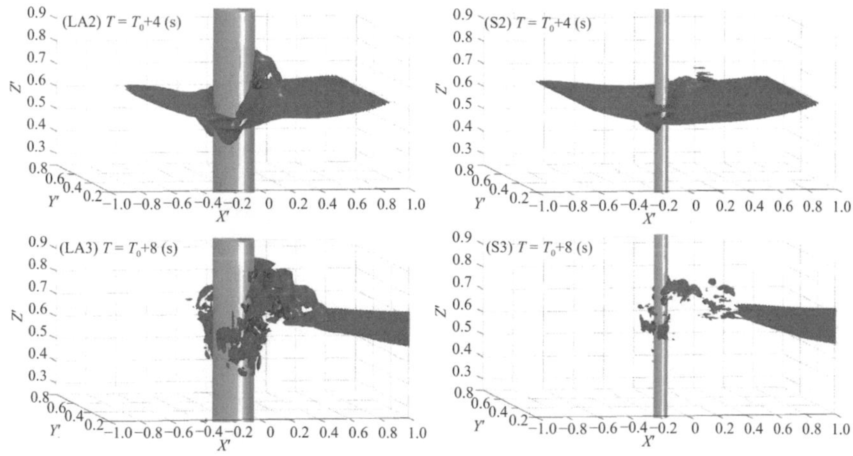

Fig. 1.8 Changes in vorticity isosurfaces during the interaction between a vertical cylinder and internal solitary waves [93]

1.2.5 Research on Interaction Between Internal Solitary Waves and Submersibles

Internal solitary waves not only affect underwater structures but also pose a threat to the safe navigation of submerged bodies. When a slender body of revolution, such as a submarine, hovers underwater, the flow field induced by internal solitary waves can exert significant sudden horizontal loads on it, potentially causing large horizontal drift motions. On the other hand, when a submarine traverses a density jump layer, the difference in density between the upper and lower layers can cause a sudden change in buoyancy. Combined with the opposite horizontal flow velocities in the upper and lower layers, this can create a shearing effect, exerting sudden horizontal shear loads on the submarine. This may lead to abrupt changes in the motion attitude of the submersible, or even loss of control, resulting in a "falling deep" phenomenon. If not controlled in time, this can be fatal to the submersible.

To investigate the impact of internal solitary waves on the forces and motion of submerged bodies, scholars first focused on the load characteristics of fixed submerged bodies under the action of internal solitary waves. Chen et al. [94], based on the KdV-mKdV theoretical solution for internal solitary waves in a two-layer fluid, used steady flow to replace the submarine's velocity to simulate the interaction between internal solitary waves and a moving submarine. They obtained the load characteristics of the moving submarine, paving a new way for future research on the interaction between internal waves and moving submerged bodies. In addition to the common downward concave type, internal solitary waves can also be upward convex when the upper layer is thicker than the lower layer. Guan et al. [95] applied the dual-paddle wave generation method to simulate a numerical wave tank for internal solitary waves and used a finite volume adaptive semi-structured multi-grid method to study the load effects of upward and downward internal solitary waves on submersibles in the ocean. They obtained the flow field morphology and load characteristics of submersibles when encountering internal solitary waves in the ocean. Moreover, the position and attitude of the submerged body also affect its load characteristics. Wu [96] et al. respectively studied the effects of amplitude and submersion depth on the horizontal and vertical forces on a horizontally placed and head-on wave-facing drop-shaped body of revolution. Gu [97–98] et al. further investigated the effects of the evolution of internal solitary waves over a ridge on the horizontal and vertical forces on a head-on wave-facing drop-shaped body of revolution. Wei [99] et al. studied the characteristics of internal solitary waves in stratified fluids with continuous density jump layers and their interaction with submerged slender bodies, and established similarity criteria for full-scale and model-scale submarines. Wang [100] used various experimental techniques to measure the internal solitary wave flow field and the forces on the submerged body, revealing that the horizontal force on the submerged body depends on the velocity field caused by internal solitary waves, while the vertical force is determined by both the velocity and density fields, as shown in Fig. 1.9a. Ji [101] et al. experimentally and theoretically studied the wave load characteristics of internal solitary waves acting on an underwater slender body. The results showed that different incident angles, wave amplitudes, and ratios of upper and lower layer fluid thicknesses have a significant impact on the wave load characteristics of the slender body, especially the internal solitary waves acting in the transverse direction, which have a more severe effect on the slender body. Based on the interaction between internal solitary waves and fixed submerged bodies, more scholars have focused on the loaded motion of free-moving submerged bodies. Du [102] et al. conducted motion response experiments using a flexible rope to suspend a submarine model. Their results emphasized that the trajectory of the submarine's center of gravity under the action of internal solitary waves is an irregular elliptical line, and the submarine's heave, sway, and pitch responses are significant. However, the motion response was affected by the constraints of the flexible rope. Cui [103] et al. studied the dynamic response characteristics of a suspended submerged body under the action of internal solitary waves and found that the wave amplitude greatly affects the heave characteristics of the submerged body, while the sway characteristics of the submerged body are influenced by the distance between the submerged body and the density jump layer. Huang [104] et al. established a numerical simulation method for the interaction between internal solitary waves in stratified flow and submersibles based on the RANS method and KdV theory. They found that internal solitary waves can make the flow field around the vehicle extremely complex, thus causing abrupt changes in fluid dynamics and motion, which should be given great attention to navigation safety at this time. Li [105] et al. developed a numerical model of three-dimensional Navier–Stokes equations with a modified k-ω SST turbulence model combined with a density transport equation to simulate the interaction of ISWs and submarines in continuously stratified, incompressible and viscous fluid, as shown in Fig. 1.9b.

Current research mainly focuses on the load characteristics of fixed marine structures under the action of internal solitary waves. The motion response characteristics

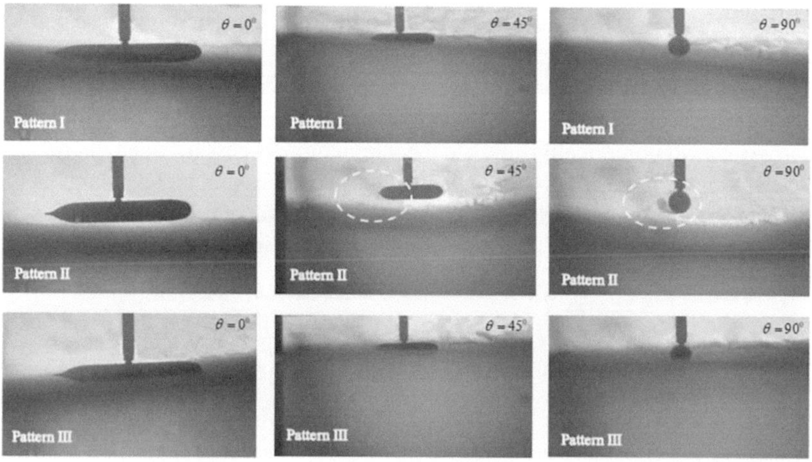

(a) Interaction between a submerged body and oblique internal solitary waves at different wave angles [100]

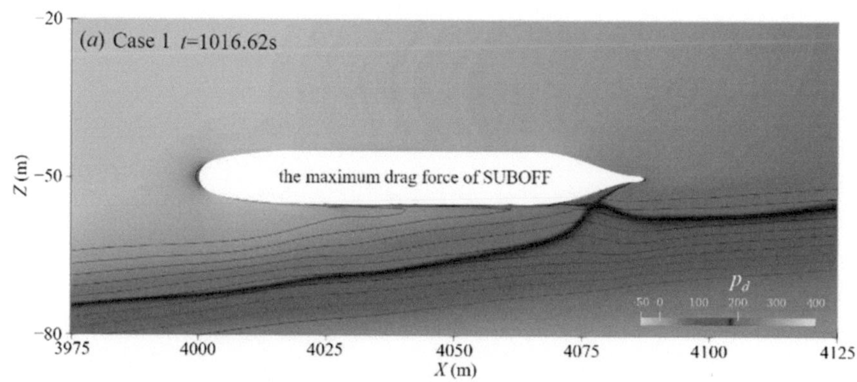

(b) Dynamic pressure field around the SUBOFF [105]

Fig. 1.9 Interaction between internal solitary waves and submersibles

of suspended submerged bodies have attracted increasing attention from scholars. However, it is challenging to maintain the suspension of the submerged body in experiments, and there is a lack of corresponding simulation data for comparison and verification. Many works have already conducted extensive simulation work on this topic, revealing that the submersion depth of the submerged body and the amplitude of internal solitary waves are important influencing factors. However, in-depth exploration of the mechanism model should exclude the influence of dimensions and not only consider the time-varying characteristics of the submerged body's load and motion. Moreover, the scale of the reduced model and internal solitary waves cannot correspond to the actual situation, resulting in research findings that lack universality.

In summary, there is still no unified understanding of the motion response and load mechanism of freely moving submerged bodies affected by internal solitary waves, and there is considerable controversy over the mechanism model. In view of this, this work establishes a three-dimensional full-scale numerical wave tank for internal solitary waves to numerically simulate the interaction process between internal solitary waves and submerged bodies in the real marine environment, revealing the laws of motion response and load characteristics of underwater submerged bodies, and providing a forecasting basis for the safe underwater navigation of submerged bodies.

Chapter 2
Internal Wave Theory

2.1 Theoretical Models of Internal Solitary Waves

When studying the propagation and evolution of internal solitary waves in stratified fluids, seawater is typically simplified as two homogeneous, immiscible fluid layers. Existing two-layer theoretical models are based on potential flow theory, assuming irrotational, incompressible flow with upper and lower layer depths set as h_1 and h_2, fluid densities set as ρ_1 and ρ_2, with total water depth $H = h_1 + h_2$. Since the density difference between upper and lower layers in actual oceans is small, the vertical velocity induced at the free surface during internal solitary wave propagation can be essentially neglected. Therefore, a rigid-lid condition can be applied at the top of the fluid domain, while the bottom is set as a rigid no-slip boundary. Taking the interface as the x axis and the vertical direction as the z axis, a Cartesian coordinate system xoz is established, as shown in Fig. 2.1.

Let a and λ represent the amplitude and characteristic wavelength of the internal solitary wave respectively [33], where the characteristic wavelength λ is defined by the following equation. The magnitude of S equals half the area enclosed between the wave surface and the x axis, which is equal to the area of the shaded rectangular region on the left. The nonlinear parameter $\varepsilon = \frac{|a|}{H}$ and dispersion parameter $\mu = (H/\lambda)^2$ are defined [36, 58], with the interface displacement of the internal solitary wave given by $\zeta(x, t)$.

$$\lambda = \frac{S}{a} = \frac{1}{2a}\int_{-\infty}^{\infty} \zeta(x, t)dx \tag{2.1}$$

However, real oceanic environments possess density transition layers of certain thickness, making the two-layer model incapable of fully characterizing the true properties of internal solitary waves. To accurately describe internal solitary waves with continuous oceanic stratification and relatively large amplitudes compared to the total water depth, this study employs the DJL (Dubreil-Jacotin-Long) equation,

P. Du et al., *Internal Wave Hydrodynamics on Submersibles*,
https://doi.org/10.1007/978-981-95-4479-0_2

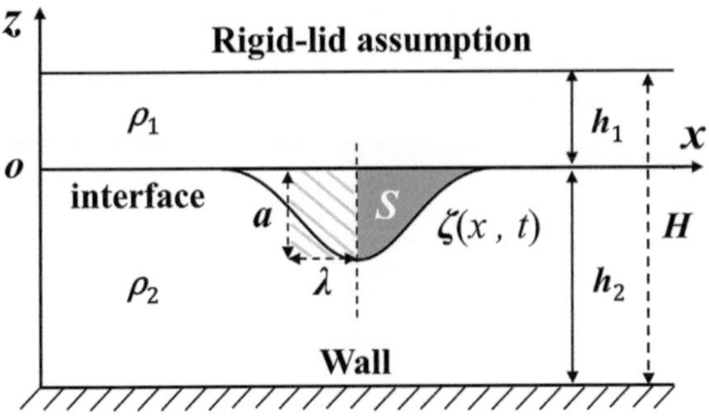

Fig. 2.1 Internal solitary wave in a two-layer model

which is equivalent to the complete set of steady-state stratified Euler equations. Since the derivation of the DJL equation makes no assumptions about the nonlinearity of fluid flow, the obtained solutions represent exact internal solitary wave solutions under inviscid conditions [41].

Different internal solitary wave theories have varying ranges of applicability. Selecting the most suitable internal solitary wave theory is a crucial step in numerical wave generation. The following sections provide detailed introductions to these theories.

2.1.1 KdV Theoretical Model

Based on the assumptions of weak nonlinearity, weak dispersion, and their balance ($\varepsilon = O(\mu) << 1$), considering a steady internal solitary wave with amplitude propagating rightward along the ox axis, the KdV equation is:

$$\zeta_t + c_0\zeta_x + c_1\zeta\zeta_x + c_2\zeta_{xxx} = 0 \tag{2.2}$$

where:

$$c_0^2 = \frac{gh_1h_2(\rho_2 - \rho_1)}{\rho_1h_2 + \rho_2h_1} \tag{2.3}$$

$$c_1 = -\frac{3c_0}{2}\frac{\rho_1h_2^2 - \rho_2h_1^2}{\rho_1h_1h_2^2 + \rho_2h_2h_1^2} \tag{2.4}$$

$$c_2 = \frac{c_0}{6}\frac{\rho_1h_1^2h_2 + \rho_2h_2^2h_1}{\rho_1h_2 + \rho_2h_1} \tag{2.5}$$

The interface displacement $\zeta(x, t)$ solution is:

$$\zeta(x, t) = \pm a \cdot sech^2(\lambda_{KdV}(x - c_{KdV}t)) \tag{2.6}$$

$$c_{KdV} = c_0 + \frac{ac_1}{3} \tag{2.7}$$

$$\lambda_{KdV} = \sqrt{\frac{ac_1}{12c_2}} \tag{2.8}$$

where c_{KdV} is the phase velocity of the wave, λ_{KdV} is the characteristic wavelength of the theoretical solution, and g is the gravitational acceleration.

The classical KdV theory is only applicable to weakly nonlinear, small-amplitude internal solitary waves. The main reason is that the KdV theory derivation requires a balance between nonlinearity and dispersion. However, when wave amplitude increases, higher-order nonlinear terms become dominant, disrupting this balance. Based on this, by adding a cubic nonlinear term to the KdV equation, the eKdV equation is obtained:

$$\zeta_t + \left(c_0 + c_1\zeta + c_3\zeta^2\right)\zeta_x + c_2\zeta_{xxx} = 0 \tag{2.9}$$

where c_0, c_1 and c_2 are the same as in the KdV theoretical solution, and the expression for c_3 is as follows:

$$c_3 = \frac{3c_0}{h_1^2 h_2^2}\left(\frac{7}{8}\left(\frac{\rho_1 h_2^2 - \rho_2 h_1^2}{\rho_1 h_2 + \rho_2 h_1}\right)^2 - \frac{\rho_1 h_2^3 + \rho_2 h_1^3}{\rho_1 h_2 + \rho_2 h_1}\right) \tag{2.10}$$

Stationary theoretical solution of internal solitary waves in eKdV:

$$\zeta(x,t) = \pm\frac{a}{B + (1-B)\cosh^2[\lambda_{eKdV}(x - c_{eKdV}t)]} \tag{2.11}$$

$$c_{eKdV} = c_0 + \frac{a}{3}\left(c_1 + \frac{1}{2}c_3\right) \tag{2.12}$$

$$\lambda_{eKdV}^2 = \frac{a(2c_1 + ac_3)}{24c_2} \tag{2.13}$$

$$B = \frac{-ac_3}{2c_1 + ac_3} \tag{2.14}$$

Limit amplitude of the eKdV theoretical solution:

$$a_{eKdV}^{max} = \frac{4h_1h_2(h_1 - h_2)}{h_1^2 + h_2^2 + 6h_1h_2} \tag{2.15}$$

The eKdV theory is suitable for weakly to moderately nonlinear internal solitary waves with small to medium amplitudes, and the designed amplitude must not exceed the limit amplitude. When the amplitude of an internal solitary wave approaches its limit amplitude, the waveform rapidly broadens, eventually exhibiting a peculiar "tabletop" waveform that does not match experimental observations [110].

On the other hand, the mKdV equation is applicable to cases where the wave amplitude varies from 0 to $\overline{h}$. Here, $\overline{h}$ represents the distance between the interface and the critical level h_c, and the solution for interface displacement $\zeta(x, t)$ is given by:

$$\zeta(x,t) = \pm a \cdot \frac{\text{sech}^2[(x - c_{mKCV}t)/\lambda_{mKaV}]}{1 - \kappa \tanh^2[(x - c_{mKCDV}t)/\lambda_{mKaV}]} \tag{2.16}$$

where:

$$\kappa = \begin{cases} h\prime/h\prime, \overline{h} > 0 \\ h\prime/h\prime, \overline{h} < 0 \end{cases}, h_c = \frac{h}{1 + \sqrt{\rho_1/\rho_2}}, \overline{h} = h_2 - h_c \tag{2.17}$$

$$h\prime = -\overline{h} - |\overline{h} + a| \quad h'' = -\overline{h} + |\overline{h} + a| \tag{2.18}$$

$$c_{mKdV} = c_{0m}\left[\frac{1}{\left(\frac{\overline{h}+a}{h-h_c}\right)^2}\right] \tag{2.19}$$

$$\lambda_{mKdV} = \frac{2(h - h_c)\sqrt{(h - h_c)^3 + h_c^3}}{\sqrt{3hh\prime h''}} \tag{2.20}$$

$$c_{0m} = \left(\frac{gh}{2}\left[1 - \left(1 - \frac{4(\rho_2 - \rho_1)h_c(h - h_c)}{\rho_2 h^2}\right)^{1/2}\right]\right)^{1/2} \tag{2.21}$$

The mKdV equation provides the closest match to experimental results in simulating the vertical distribution of horizontal velocities in internal solitary waves. It is applicable to various upper-to-lower layer thickness ratios and exhibits strong nonlinearity [103], showing good agreement with laboratory-scale experimental results. It has certain advantages in simulating larger-amplitude internal solitary waves, with its limit amplitude given as follows:

$$a_{mKaV}^{max} = \frac{h_1\sqrt{\rho_2/\rho_1} - h_2}{\sqrt{\rho_2/\rho_1} + 1} \tag{2.22}$$

2.1.2 MCC Theoretical Model

Choi [34] and Camassa [111] et al. derived a fully nonlinear internal solitary wave theory known as the MCC theory, which is based on the original Euler equations with the sole assumption that the wavelength of internal solitary waves is greater than the thickness of any undisturbed fluid layer. It considers full nonlinearity ($O(\varepsilon) = 1$) while retaining only weak dispersion effects ($\mu << 1$). Extensive research in stratified fluid fields has demonstrated that MCC solitary wave solutions show good agreement with laboratory experiments and observational results. According to MCC theory, a nonlinear ordinary differential equation for interface displacement can be obtained:

$$(\zeta_x)^2 = \left[\frac{3g(\rho_2 - \rho_1)}{c^2\left(\rho_1 h_1^2 - \rho_2 h_2^2\right)}\right]\frac{\zeta^2(\zeta - a_+)(\zeta - a_-)}{(\zeta - a_*)} \tag{2.23}$$

Here, a_* is expressed as:

$$a_* = -\frac{h_1 h_2(\rho_1 h_1 + \rho_2 h_2)}{\rho_1 h_1^2 + \rho_2 h_2^2} \tag{2.24}$$

$a_\pm$ is the root of $\zeta^2 + q_1\zeta + q_2 = 0$, where q_1 and q_2 are defined as follows:

$$q_1 = -\frac{c_{MCC}^2}{g} - h_1 + h_2, \quad q_2 = h_1 h_2\left(\frac{c_{MCC}^2}{c_0^2} - 1\right) \tag{2.25}$$

$$\frac{c_{MCC}^2}{c_0^2} = \frac{(h_1 - a)(h_2 + a)}{h_1 h_2 - \left(c_0^2/g\right)a} \tag{2.26}$$

In Eq. (2.26), c_0 represents the linear long-wave velocity, same as the c_0 in the KdV equation; c_{MCC} is the phase velocity of internal solitary waves.

2.1.3 DJL Theoretical Model

The DJL equation is a theory describing fully nonlinear internal solitary waves, equivalent to the complete set of Euler equations for stratified fluids. It proposes a concise numerical solution method for permanent-form translating waves, through which a steadily moving flow field at constant velocity can be computationally obtained [41]. This theory considers inviscid, incompressible, irrotational fluids under the Boussinesq approximation, and the DJL equation expression in a two-dimensional fixed reference frame without considering background flow effects is as follows:

$$\nabla^2\eta + \frac{N^2(z - \eta)}{c^2}\eta = 0 \tag{2.27}$$

$$\eta(x, 0) = \eta(x, H) = 0 \tag{2.28}$$

$$\eta \to 0, \text{as}|x| \to \infty \tag{2.29}$$

$$N^2(z) = -g\frac{d\overline{\rho}(z)}{dz} \tag{2.30}$$

where N^2 represents the square of the buoyancy frequency, η denotes the isopycnal displacement, c indicates the wave propagation speed, $\overline{\rho}(z)$ shows the vertical distribution of average seawater density, x and z represent horizontal and vertical coordinates respectively, H signifies the total water depth, $|x| \to \infty$ means the absolute value of horizontal coordinate tends to infinity, and g is the gravitational acceleration.

By solving the above DJL equation, we can obtain η and c. Then through the relationship of $\varphi = \eta c$, we can derive the wave-induced velocity field, where φ is the stream function. The computational results of the DJL equation [112] are interpolated into CFD to obtain the initial field required for numerical simulation.

2.1.4 Comparison of Internal Solitary Wave Theoretical Equations

With the upper layer fluid thickness set at 50 m and density at $\rho_1 = 1020\,\text{kg/m}^3$, lower layer fluid thickness at 250 m and density at $\rho_2 = 1028\,\text{kg/m}^3$, and maximum amplitude at 50 m, the waveforms of various internal solitary wave theoretical solutions are plotted as shown in Fig. 2.3a. The shaded area represents the continuously stratified internal solitary wave calculated by the DJL equation, featuring a pycnocline thickness of 10 m. From the waveform comparison, under current fluid parameter settings, the KdV equation yields the narrowest theoretical waveform, showing significant differences compared with other theoretical waveforms. This is because KdV is suitable for small-amplitude cases where the wave amplitude to total depth ratio is less than 0.1. The mKdV equation produces the widest waveform, consistent with conclusions from other literature [35], as mKdV is applicable to weakly and moderately nonlinear internal solitary waves, showing good agreement with laboratory experimental results [103]. The eKdV equation yields waveforms between KdV and mKdV, also suitable for weakly to moderately nonlinear internal solitary waves. The MCC and DJL equations, derived from Euler equations, represent fully nonlinear internal solitary wave theories applicable to large-amplitude waves. The MCC waveform shows good agreement with the DJL waveform at the density median $\overline{\rho} = 1024\,\text{kg/m}^3$. Ignoring amplitude errors, the MCC theoretical solution can be considered as one solution within the complete Eulerian solutions of DJL.

Figure 2.2b shows the theoretical wave speeds and characteristic wavelengths calculated by various theoretical solutions under the current fluid parameter settings. It can be observed that the KdV equation yields the highest theoretical wave speed, while the DJL equation's wave speed is lower than that of strongly stratified theories

(KdV-type equations and MCC equations). The KdV equation produces the narrowest waveform, corresponding to the smallest characteristic wavelength (calculated using Eq. (2.8)). The characteristic wavelength of the DJL equation is calculated based on the waveform at the median density level, and it is found to be nearly equal to that of the MCC equation, further indicating good agreement between the DJL equation's waveform at the median density level and the MCC equation's waveform.

Fig. 2.2 Comparison of internal solitary wave theoretical solutions

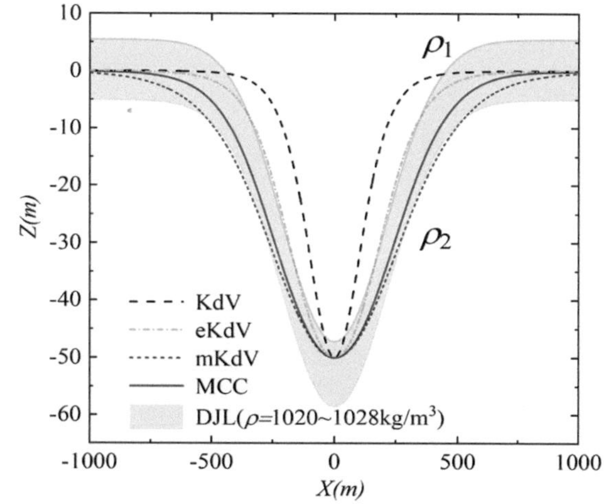

(a) Waveform comparison

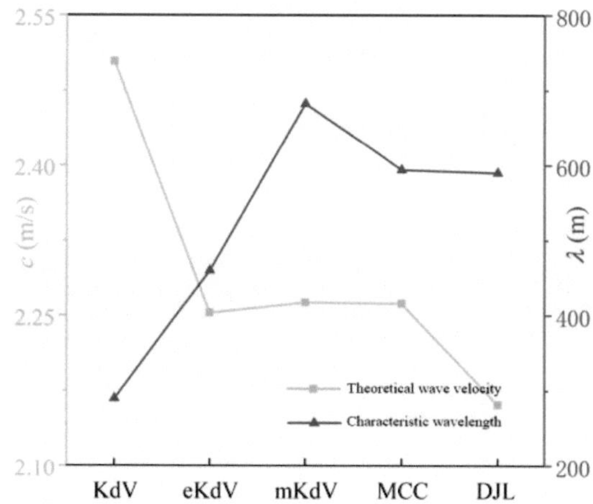

(b) Theoretical wave speed and characteristic wavelength

Chapter 3
Numerical Methods and Validations

3.1 Governing Equations of the Model

Since this study focuses on the motion and load characteristics of submerged bodies under the influence of oceanic internal solitary waves, the research process neglects temperature effects and considers only the laws of mass conservation and momentum conservation. The governing equations are as follows [106].

(1) Mass Conservation Equation (Continuity Equation)

For a fixed control volume in a Cartesian coordinate system, the net mass flow rate out of the control volume per unit time equals the time rate of mass decrease within the control volume during the same interval. This derivation yields the differential form of the continuity equation for fluid motion in Cartesian coordinates:

$$\frac{\partial \rho}{\partial t} + \frac{\partial (\rho u_i)}{\partial x_i} = 0 \tag{3.1}$$

Using the divergence formula, Eq. (2.1) can be rewritten as:

$$\frac{\partial \rho}{\partial t} + div(\rho u) = 0 \tag{3.2}$$

For incompressible flow, Eq. (2.2) can be expressed in the following vector form:

$$div(u) = 0 \,\text{or}\, \nabla \cdot u = 0 \tag{3.3}$$

P. Du et al., *Internal Wave Hydrodynamics on Submersibles*,
https://doi.org/10.1007/978-981-95-4479-0_3

(2) Momentum Conservation Equation (Navier–Stokes Equation)

The momentum equation mathematically describes the relationship between fluid motion and the forces acting upon it. Based on Newton's second law ($F = ma$), for a fluid control volume, the left side of the equation represents the sum of body forces and surface forces acting on the control volume, while the right side is the product of the control volume's mass and the material derivative of its velocity. This derivation yields the differential form of the momentum conservation equation:

$$\rho\left(\frac{\partial u_i}{\partial t} + u_j\frac{\partial u_i}{\partial x_j}\right) = \rho f_i - \frac{\partial p}{\partial x_i} + \mu\frac{\partial^2 u_i}{\partial x_j \partial x_j} \tag{3.4}$$

For incompressible viscous flow, Eq. (3.4) can be expressed in the following vector form:

$$\frac{du}{dt} = f_i - \nabla p + v\Delta u \tag{3.5}$$

In the above equations, ρ represents fluid density; u represents the velocity vector, u_i denotes the velocity component in the $i(i = 1, 2, 3)$ direction, $u = (u_1, u_2, u_3)$; p represents the pressure field; μ denotes the dynamic viscosity coefficient; f_i represents the component of body force acting on the fluid in the direction.

Considering that engineering applications are more concerned with pressure loss and average velocity distribution of flow rather than turbulence details, the Reynolds-averaged method is more widely applied. The Reynolds-averaged method [107] decomposes instantaneous values of velocity and pressure into time-averaged quantities and fluctuating quantities, establishing N-S equations that reflect time-averaged values of turbulence motion elements. While reducing computational load, it can calculate complex flows with large Reynolds numbers.

The governing equation form for incompressible, viscous fluids based on Reynolds-averaged N-S is as follows:

$$\frac{\partial \overline{u}_i}{\partial x_i} = 0 \tag{3.6}$$

$$\frac{\partial \overline{u}_i}{\partial t} + \overline{u}_j\frac{\partial \overline{u}_i}{\partial x_j} = f_i - \frac{1}{\rho}\frac{\partial \overline{p}}{\partial x_i} + \frac{\partial}{\partial x_j}\left[\nu\left(\frac{\partial \overline{u}_i}{\partial x_j} + \frac{\partial \overline{u}_j}{\partial x_i}\right)\right] - \frac{\partial \overline{u_i\prime u_j\prime}}{\partial x_j} \tag{3.7}$$

where $\overline{u}_i$ represents the time-averaged velocity of the fluid in the i direction; $\overline{p}$ is the time-averaged pressure; ν is the kinematic viscosity coefficient; $\overline{u_i\prime u_j\prime}$ is the Reynolds stress term.

3.2 Numerical Methods

3.2.1 Turbulence Model

The existence of Reynolds stress terms makes RANS equations unclosed, therefore turbulence models are constructed to close the RANS equations. In practical engineering fields, the Standard $k-\varepsilon$ model is typically used to close the above equation system, where the equations for turbulent kinetic energy k and turbulent dissipation rate ε are respectively:

$$\rho\left[\frac{\partial k}{\partial t}+\frac{\partial(k\overline{u}_i)}{\partial x_i}\right]=\frac{\partial}{\partial x_j}\left[\left(\mu+\frac{\mu_t}{\sigma_k}\right)\frac{\partial k}{\partial x_j}\right]+G_k-\rho\varepsilon \tag{3.8}$$

$$\rho\left[\frac{\partial \varepsilon}{\partial t}+\frac{\partial(\varepsilon\overline{u}_i)}{\partial x_i}\right]=\frac{\partial}{\partial x_j}\left[\left(\mu+\frac{\mu_t}{\sigma_\varepsilon}\right)\frac{\partial \varepsilon}{\partial x_j}\right]+\frac{C_{1\varepsilon}\varepsilon}{k}G_k-C_{2\varepsilon}\rho\frac{\varepsilon^2}{k} \tag{3.9}$$

Turbulent viscosity equation:

$$\mu_t=\rho C_\mu\frac{k^2}{\varepsilon} \tag{3.10}$$

where: $\overline{u}_i$ is the average turbulent velocity; μ_t is the turbulent viscosity coefficient; G_k is the production term of turbulent kinetic energy k caused by mean velocity gradient, σ_k and σ_ε are the Prandtl numbers corresponding to turbulent kinetic energy k and dissipation rate ε respectively, $C_{1\varepsilon}$ and $C_{2\varepsilon}$ are empirical constants. In actual calculations, these coefficients are often taken as: $C_{1\varepsilon}=1.44$, $C_{2\varepsilon}=1.92$, $C_\mu=0.09$, $\sigma_k=1.0$, $\sigma_\varepsilon=1.3$.

The Shear Stress Transport (SST) model is an excellent all-round RANS model, and currently the best-performing one among widely used eddy viscosity assumption models, thus highly regarded by many scholars [108]. The standard SST $k-\omega$ turbulence model mainly consists of two transport equations for turbulent kinetic energy k and specific turbulent dissipation rate ω, with expressions as follows:

$$\frac{\partial}{\partial t}(\rho k)+\frac{\partial}{\partial x_i}(\rho k\overline{u}_i)=\frac{\partial}{\partial x_i}\left(\Gamma_k\frac{\partial k}{\partial x_j}\right)+G_k-Y_k+S_k \tag{3.11}$$

$$\frac{\partial}{\partial t}(\rho\omega)+\frac{\partial}{\partial x_i}(\rho\omega\overline{u}_i)=\frac{\partial}{\partial x_i}\left(\Gamma_\omega\frac{\partial \omega}{\partial x_j}\right)+G_\omega-Y_\omega+D_\omega+S_\omega \tag{3.12}$$

where: Γ_k, Γ_ω are effective diffusion coefficients, G_k and G_ω are turbulence production terms; Y_k and Y_ω are dissipation terms for k and ω; D_ω is the diffusion term.

The near-wall region can be roughly subdivided into three layers. The layer closest to the wall is called the "viscous sublayer," where the flow is nearly laminar and

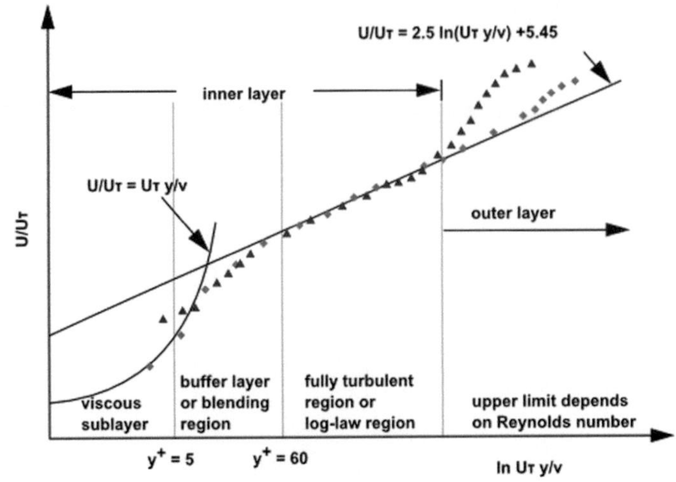

Fig. 3.1 Division of the near-wall region [109]

viscosity plays a dominant role in momentum, heat, and mass transfer. The outer layer is known as the fully turbulent layer, where turbulence predominates, and the flow is in a fully developed turbulent state with a velocity profile approximating the logarithmic law, hence also called the logarithmic law layer. Between the viscous sublayer and the fully turbulent layer lies a transition region called the buffer layer, as shown in Fig. 3.1.

Generally, to simplify grid generation and improve computational efficiency, the wall function method is used to solve the near-wall region. y^+ (dimensionless wall distance) serves as the criterion for selecting the solution method in the near-wall region, and the expression for y^+ is as follows:

$$y^+ = \frac{y\rho u_\tau}{\mu} = \frac{y}{v}\sqrt{\frac{\tau_w}{\rho}} \tag{3.13}$$

where τ_w is the wall shear stress, y is the normal distance to the wall, ρ is the fluid density, and μ is the dynamic viscosity coefficient.

Different turbulence models correspond to different values of y^+. For the $k - \varepsilon$ turbulence model, the range of y^+ is 30–300; for the SST $k - \omega$ turbulence model, y^+ is around 1. In high-Reynolds-number turbulent flows, the Standard $k - \varepsilon$ model is more widely used in engineering applications, as its convergence and computational accuracy meet general engineering requirements. However, it performs poorly in simulating complex flows with large curvatures or strong pressure gradients. On the other hand, the SST $k - \omega$ turbulence model is highly dependent on wall distance, making it unsuitable for free shear flows and requiring higher grid quality. Therefore, this study selects the Standard $k - \varepsilon$ turbulence model.

3.2.2 Multiphase Flow Model

To simulate internal solitary waves in stratified fluids, a multiphase flow model is typically required for flow field solutions. Currently, the commonly used multiphase flow research method is the Euler-Euler approach. The Fluent solver provides two Euler-Euler multiphase flow models: VOF (Volume Of Fluid) and the Mixture model, which are described in detail below.

(1) VOF Model

The VOF model is a surface-tracking method based on a fixed Eulerian grid. During computation, the volume fraction α is introduced to describe the proportion and density of different phases. Internal solitary waves form at the interface between two fluid layers, so tracking the two-phase interface is necessary. The VOF (Volume Of Fluid) method is used to track changes in the interface between the two fluid layers. The fluid volume function $\alpha(z)$ is defined to mark the state of the two fluids in each grid cell. Here, α_i represents the volume fraction of the i-th phase fluid, $\alpha_i = 0$ indicates the absence of the i-th phase fluid in the cell, and $\alpha_i = 1$ indicates that only the i-th phase fluid exists in the cell.

$$\alpha_i(z) = \begin{cases} 0 & \text{absence of the } i - th \text{ phase} \\ 0 < \alpha < 1 & \text{interface} \\ 1 & \text{the i} - \text{th phase} \end{cases} \tag{3.14}$$

(2) Mixture Model

The Mixture model is a simplified multiphase flow model with broad applications. It can simulate multiphase flows where phases move at different velocities, and considers mass exchange more comprehensively than the VOF model. When simulating continuously stratified internal solitary waves, it's necessary to account for density variations at the pycnocline and mass/momentum exchanges between phases. Notably, the Mixture model can only be used for pressure-based solvers, not density-based ones, and viscous effects must be considered during calculations.

In summary, the VOF model is suitable for stratified flows and free surface flows, while the Mixture and Eulerian models are better for flows with phase mixing or separation. For simpler problems with lower accuracy requirements, the Mixture model is preferable. Therefore, we select the VOF model for simulating strongly stratified internal solitary waves, and the Mixture model for continuously stratified cases.

3.2.3 *Dynamic Mesh Technology*

The dynamic mesh model can simulate time-dependent changes in domain boundaries. Its implementation requires initial mesh generation and motion descriptions for moving zones. In Fluent, motion can be defined through boundary profiles (Boundary profiles), user-defined functions (UDFs), or six-degree-of-freedom (6DOF solver) solvers. The volume mesh positions are automatically updated each time step based on new boundary positions.

When simulating the motion response of submerged bodies under internal solitary waves, dynamic mesh technology is required to handle mesh movement/deformation. Common techniques include smoothing, layering, remeshing, and overset grid methods. This section focuses on the overset grid approach.

The overset grid divides the computational domain into a background grid and one (or multiple) component grids, with overlapping regions between them where data exchange occurs through interpolation. The workflow consists of three parts: a. Hole-cutting—identifies cells outside the computational domain and eliminates redundant overlapping cells to simplify solving. b. Establishing domain connectivity—first determines boundary points through given overset conditions, then identifies overlapping regions between background and component grids, finally matching active computational cells. c. Interpolation—establishes data transfer between boundary points and computational cells for interpolation calculations.

Since the overset grid method doesn't involve mesh deformation or reconstruction, it offers advantages over traditional deforming mesh methods for simulating transient large-amplitude motions. Therefore, this study employs the overset grid method for submerged body motion simulation.

3.3 Numerical Wave Generation Model

3.3.1 *Strongly Stratified Wave Generation and Result Analysis*

A three-dimensional strongly stratified internal solitary wave numerical flume is established as shown in Fig. 3.2. The flume has a length of 5000 m, a width of 200 m, and a height of 300 m. Since this chapter focuses on the interaction between strongly stratified internal solitary waves and submerged bodies, the thickness of the density interface is neglected when simulating internal solitary waves. The upper and lower layers of fluid are treated as two immiscible liquids, where the upper layer has a thickness of $h_1 = 50\,\text{m}$ and a density of $\rho_1 = 1020\,\text{kg/m}^3$; the lower layer has a thickness of $h_2 = 250\,\text{m}$ and a density of $\rho_2 = 1028\,\text{kg/m}^3$. The wave amplitude is set to $a = 50\,\text{m}$, and the dynamic viscosity of the fluid is $\mu = 0.001\,\text{Ns/m}^2$.

The velocity-inlet wave-making method is employed to generate strongly stratified internal solitary waves numerically. The average horizontal velocities U_1 and U_2 of

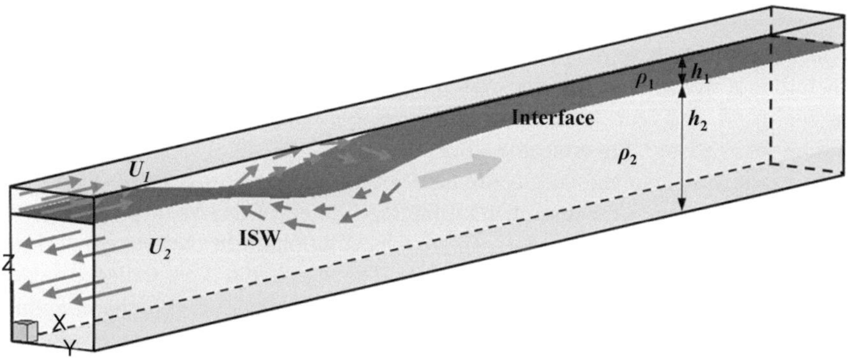

Fig. 3.2 Three-dimensional strongly stratified internal solitary wave numerical flume

the upper and lower fluid layers induced during the propagation of internal solitary waves are calculated using Eq. (3.15) [125]. The velocity control equation is then applied to the inlet boundary through a UDF (User-Defined Function).

$$U_1 = -c_{MCC} \cdot \frac{\zeta}{h_1 - \zeta}, U_2 = c_{MCC} \cdot \frac{\zeta}{h_2 + \zeta} \tag{3.15}$$

In the above equation: c_{MCC} represents the phase velocity of the MCC theoretical solution, and ζ denotes the wave surface displacement.

The ICEM software was used to perform structured meshing of the water tank model, with grid refinement applied to the path swept by internal solitary wave propagation. It was also ensured that the grid size in the action zone and the grid near the submerged body boundary could transition smoothly, thereby guaranteeing higher accuracy in data interaction between the moving submerged body and the background grid. Additionally, the grid in the wave-absorbing zone was coarsened to enhance dissipation. After meshing, the total number of grids in the water tank was approximately 15 million. A front view of the water tank grid cross-section is shown in Fig. 3.3.

The left end of the numerical water tank was set as a velocity inlet (Velocity-inlet), while the right end was set as a pressure outlet (Pressure-outlet). Considering the

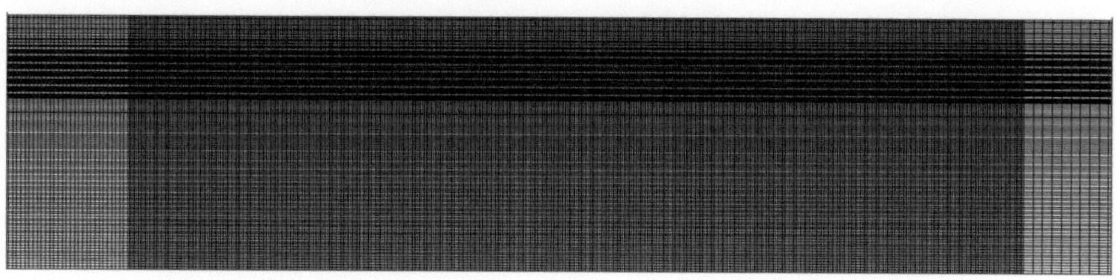

Fig. 3.3 Grid refinement of the numerical water tank

minor density variation between the upper and lower fluid layers, the vertical velocity induced by internal solitary wave propagation at the free surface was negligible. Therefore, a rigid-lid assumption was applied to the top of the fluid domain, with the top, front, and rear walls all set as symmetry boundaries (Symmetry), and the bottom set as a rigid, impermeable solid wall (Wall).

The numerical solver was configured as an unsteady transient solver. The governing equations were discretized using the Finite Volume Method (FVM), and the VOF method was employed to capture the evolution of the two-phase interface, with the volume fraction set to explicit mode. The Standard $k - \varepsilon$ turbulence model, with first-order implicit time discretization, PISO for pressure–velocity coupling, least squares method for node pressure gradient calculation, volume force-weighted pressure interpolation scheme, second-order upwind scheme for momentum equations, convection terms, and turbulence transport terms, and a tentative time step of 0.1 s.

The calculated horizontal velocity field of strongly stratified internal solitary waves is shown in Fig. 3.4a, and b shows the streamline diagram. It can be observed that the horizontal velocity is greater at the central trough of the strongly stratified internal solitary wave. The upper layer fluid reaches its maximum positive velocity at the trough position, while the lower layer fluid reaches its maximum negative velocity below the trough. Near the wave surface, the horizontal velocity transitions from positive to negative, and there is a region with zero horizontal velocity beneath the trailing wave. The streamlines bend at the left and right wave surfaces, and the flow field is asymmetric.

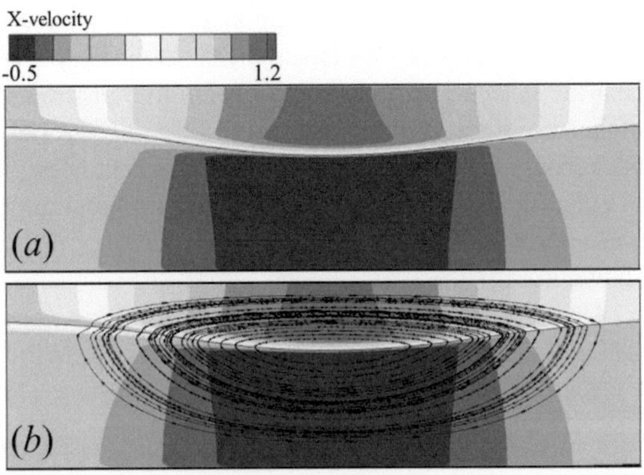

Fig. 3.4 Horizontal velocity distribution of strongly stratified internal solitary waves **a** Horizontal velocity field **b** Streamline diagram

3.4 Continuous Stratification Wave Generation and Result Analysis

Both the KdV equation and the MCC equation are based on the two-layer model theory, where the densities of the upper and lower layers are uniform and do not mix, ignoring the thickness of the pycnocline. However, in actual marine environments, there is a thickness to the pycnocline, and the two-layer model cannot fully characterize the true properties of internal solitary waves. The DJL equation is a theory that satisfies continuous stratification and large-amplitude nonlinear internal solitary waves in the ocean. Below, we use the initial flow field method to achieve numerical wave generation for continuously stratified internal solitary waves.

A three-dimensional numerical wave tank for continuously stratified internal solitary waves is established as shown in Fig. 3.5. The tank dimensions and grid are consistent with those of the strongly stratified internal solitary wave numerical tank. For simulating continuous internal solitary waves, the wave amplitude is first set to 50 m, the pycnocline thickness to 20 m, and the center of the pycnocline is positioned 50 m below the upper surface of the tank. The thickness and density of the upper and lower layers are the same as those in the strongly stratified internal solitary wave setup. By solving the DJL equation [112], a two-dimensional initial flow field is obtained, with dimensions of 2000 $\times$ 300 m (as shown in Fig. 3.5a). Since the generated internal solitary wave can only be located at the center of the computational domain, the flow field is extended longitudinally to obtain a new two-dimensional flow field with dimensions of 5000 $\times$ 300 m (as shown in Fig. 3.4b). Further, the new two-dimensional flow field is replicated and extended in the spanwise direction to form a three-dimensional flow field (as shown in Fig. 3.4c). Finally, the flow field is interpolated into the tank grid to complete the numerical wave generation for continuous density stratification (as shown in Fig. 3.4d).

The bottom of the numerical wave tank is set as a rigid, impermeable wall (Wall), while the other boundaries are set as symmetric boundaries (Symmetry). The numerical solver is configured as an unsteady transient solver, and the governing equations are discretized using the Finite Volume Method (FVM). The Mixture multiphase flow model is employed to account for mass and momentum exchange at the phase interface, and the volume fraction is set to implicit mode. The Standard $k - \varepsilon$ turbulence model was selected. The temporal discretization adopted a first-order implicit scheme, the pressure–velocity coupling was set to PISO, the node pressure gradient calculation used the least squares method, and the pressure interpolation scheme selected the body-force-weighted method. Similarly, the momentum equation, convection term, and turbulence transport term were all set to second-order upwind scheme, with a tentative time step of 0.1 s.

The computed horizontal velocity field of the continuously stratified internal solitary wave is shown in Fig. 3.6a, and the streamline diagram is shown in Fig. 3.6b. The vertical distribution of the horizontal velocity of the continuously stratified internal solitary wave transitions smoothly at the pycnocline. The flow field exhibits perfect symmetry about the vertical centerline of the wave trough, with smooth and curved

Fig. 3.5 Numerical wave generation process for continuously stratified internal solitary waves

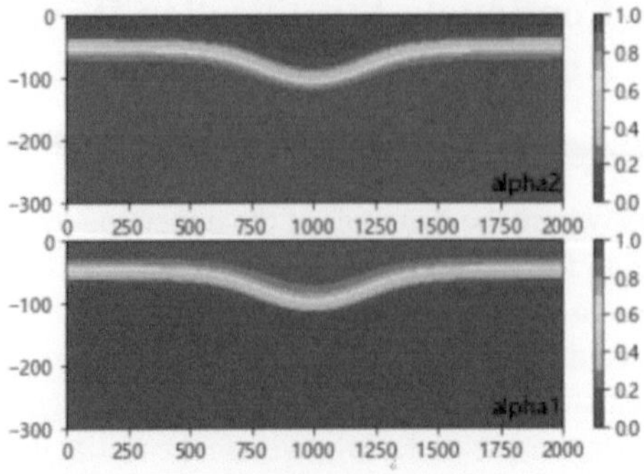

(a) Two-dimensional initial flow field

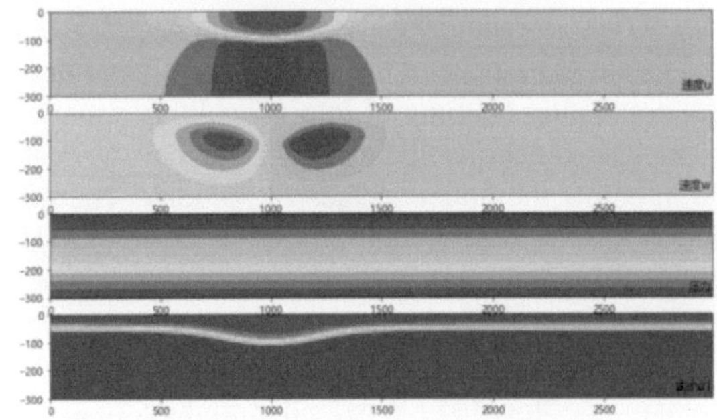

(b) Extended two-dimensional flow field

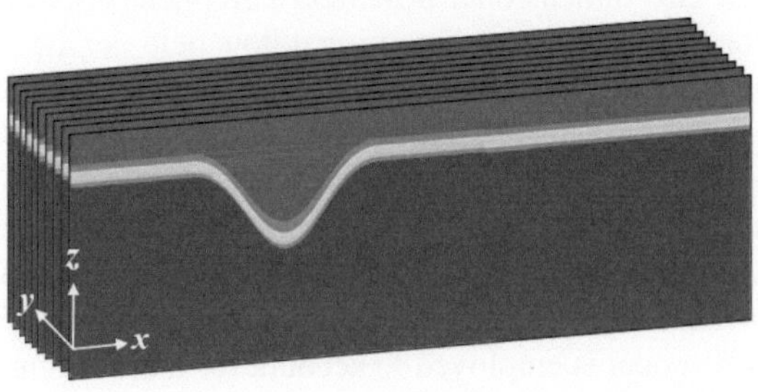

(c) Three-dimensional initial flow field

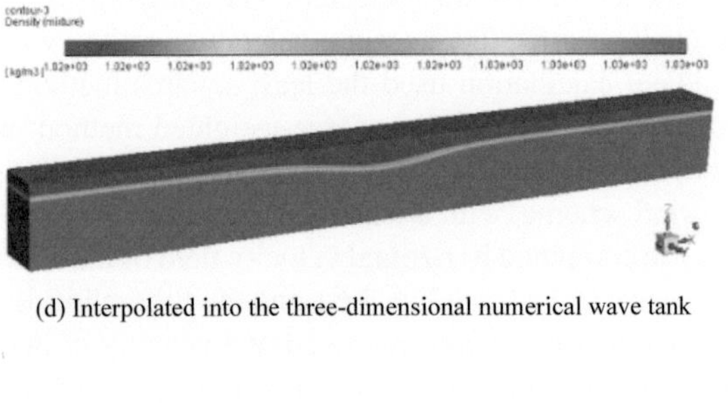

(d) Interpolated into the three-dimensional numerical wave tank

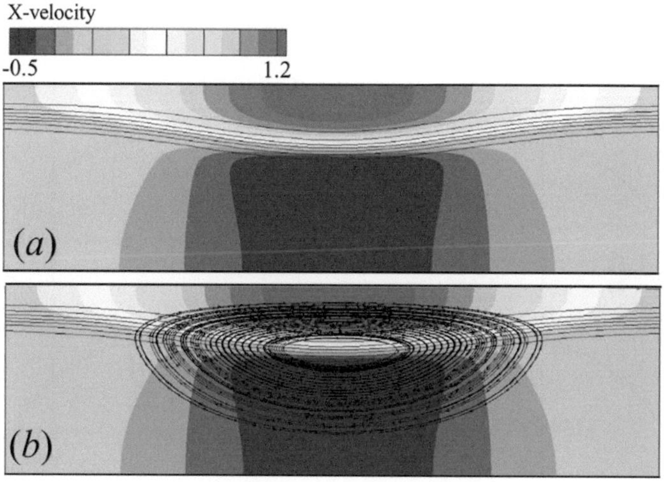

Fig. 3.6 Horizontal velocity distribution of continuously stratified internal solitary wave **a** Horizontal velocity field **b** Streamline diagram

streamlines, showing no influence of trailing waves. The horizontal velocity near the median density is approximately zero.

3.5 Verification of Independence from Wave Generation Time Step

In the numerical wave generation process, the size of the wave generation time step directly affects whether the internal solitary wave propagates stably. Therefore, it is necessary to verify the independence from the time step. In addition to the baseline time step of 0.1 s, $\Delta t = 0.05$ s and $\Delta t = 0.2$ s were also set. The comparison between simulation results and theoretical solutions is shown in Fig. 3.7.

From Fig. 3.7a, it can be observed that the horizontal velocity of the MCC theoretical solution exhibits a sudden jump at the interface, transitioning directly from positive to negative values. The simulation results for the three time steps are similar, with the flow velocities in the upper and lower layers matching well with the MCC theoretical solution. The change in horizontal velocity at the interface is more gradual. Figure 3.7b shows that the horizontal velocity of the continuously stratified internal solitary wave varies linearly at the pycnocline, with a smooth transition between the upper and lower layers, closely matching the DJL theoretical solution. The simulation results for the three different time steps are almost identical to the theoretical solutions. Considering both computational accuracy and efficiency, a wave generation time step of $\Delta t = 0.1$ s is sufficient.

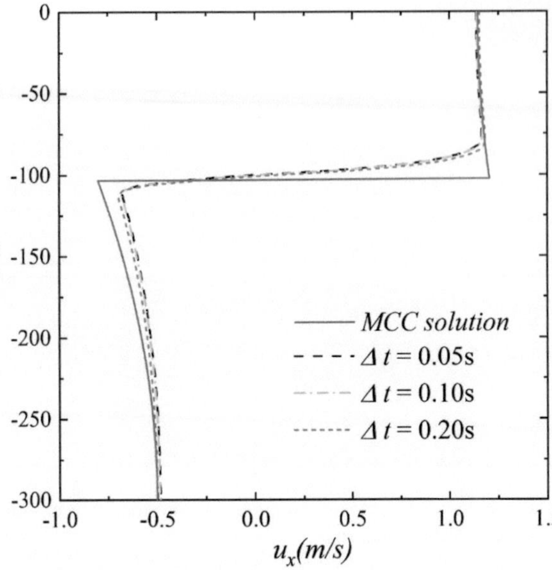

(a)Vertical distribution of horizontal velocity for strongly stratified internal solitary wave

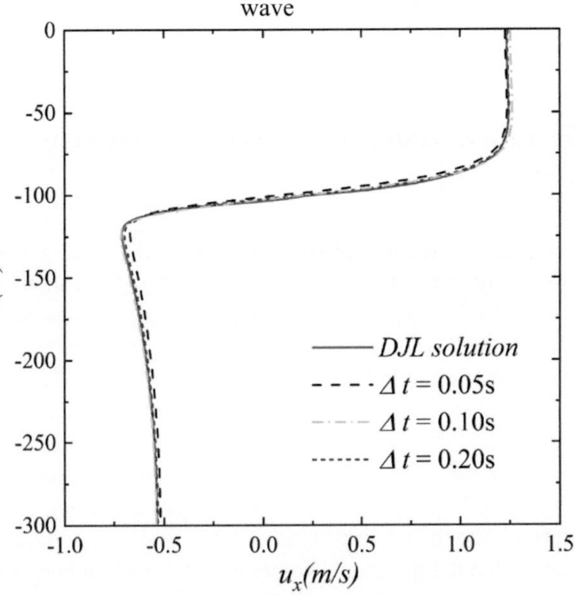

(b) Vertical distribution of horizontal velocity for continuously stratified internal solitary wave

Fig. 3.7 Verification of independence from wave generation time step

3.6 Attenuation Law of Internal Solitary Wave Amplitude

Due to numerical dissipation and error accumulation, the amplitude of internal solitary waves attenuates during propagation. By extracting the wave profiles at different times and locating the lowest point of the wave trough, the variation of amplitude with time is plotted in Fig. 3.8. The dashed line represents the linear fitting result, indicating that the amplitude attenuation over time can be approximately regarded as linear. Based on this attenuation law, the initial input wave amplitude can be increased to ensure that, when the internal solitary wave interacts with the submerged body, the amplitude has decayed to the specified value.

Moreover, as can be seen from Fig. 3.8, the attenuation rate of strongly stratified internal solitary waves is greater than that of continuously stratified internal solitary waves. The main reason is that strongly stratified internal solitary waves ignore the thickness of the density transition layer, resulting in faster propagation speed and wave height attenuation compared to continuously stratified internal solitary waves [72].

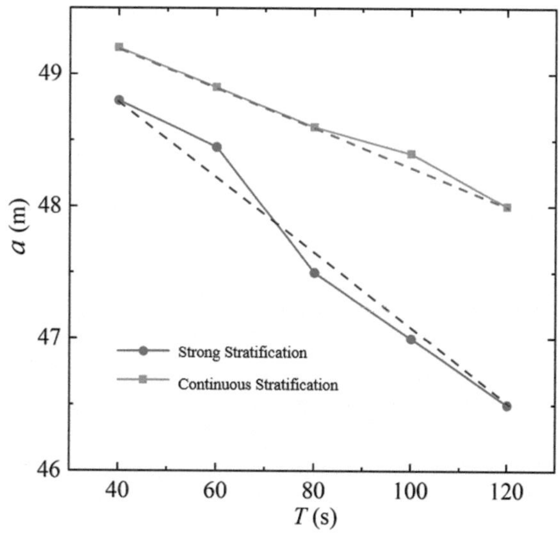

Fig. 3.8 Attenuation characteristics of internal solitary waves

Table 3.1 Geometric parameters of SUBOFF model (scale ratio 20:1)

Model scale	Length L (m)	Diameter D (m)	Displaced volume VS(m^3)	Wetted surface area $S1$ (m^2)	Maximum axial cross-sectional Area $S2$ (m^2)
Original scale	4.356	0.508	0.71	6.34	0.22
Actual scale	87.12	10.16	5645.34	2535.71	86.12

3.7 Submersible Model and Grid Independence Verification

3.7.1 Computational Model of Submersible

In terms of model selection, to ensure computational efficiency and save computing time, many scholars have chosen simplified ellipsoids, cylinders, or scaled SUBOFF bare hulls. To better approximate the real scale, this paper selects a 20-times enlarged SUBOFF-AFF8 fully appended submarine model to simulate an actual-scale submarine. The appendages include the sail and stern planes, with the stern plane cross-section being NACA0020 airfoil, symmetrically arranged at the stern. The horizontal position of the submersible's center of gravity is 40.18 m from the bow. The geometric parameters of the original SUBOFF model and the 20-times enlarged submarine model are shown in Table 3.1.

The grid of the submersible model significantly affects the accuracy of load simulation on the submersible. The boundary layer is generally divided along the wall-normal direction into viscous sublayer, buffer layer, and log-law layer. In these layers, the flow in the log-law layer is in a fully developed turbulent state, with velocity distribution approximating the logarithmic law. To study the interaction between internal solitary waves and the SUBOFF submarine, we considered fully developed turbulence, so the first grid control point should be located in the log-law region of the boundary layer. The boundary layer region division can be determined by the dimensionless value y^+. Local grid refinement in ICEM was used to divide the submersible's boundary layer grid, ensuring the first grid control point is in the log-law region. Twenty-five layers of grid were used as transition on the SUBOFF surface, with a grid size increase ratio of 1.2 and minimum grid size of 0.017 m. The structured grid of the SUBOFF submarine is shown in Fig. 3.9.

3.7.2 Grid Independence Verification

To verify the grid independence of the SUBOFF submarine, three sets of grids were generated. To ensure the grid topology remained unchanged, the total number of 3D grids was increased by a factor of $\sqrt{2}$. It is worth noting that grid refinement should be comprehensive (global) to verify independence in all directions, and the first layer of wall-adjacent grids must remain within a reasonable range during refinement. The

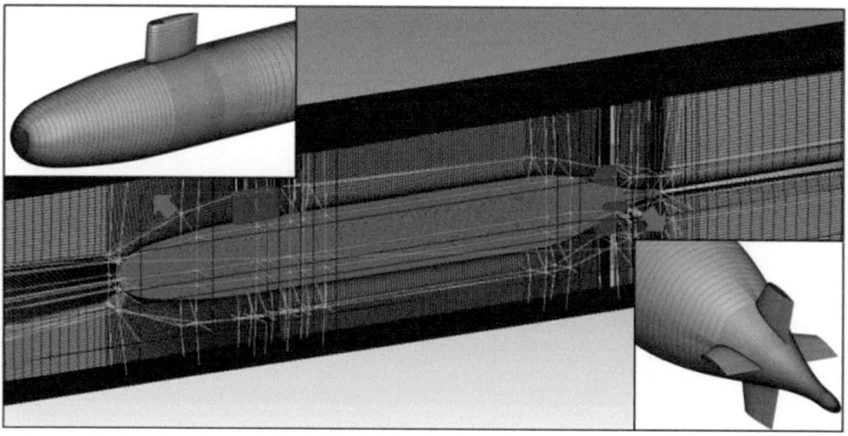

Fig. 3.9 Structured grid division of SUBOFF submarine

three grid sets, from coarse to fine, are: 3 million (Coarse), 5 million (Medium), and 7 million (Fine). The background grids were also increased accordingly to ensure consistent transitions in overlapping regions.

The SUBOFF submarine model was placed at the center of the flow field. The boundary conditions were set as follows: Symmetry on all sides, velocity inlet for the inflow surface, and pressure outlet for the outflow surface. The submarine's bow was positioned 1.5 times the hull length from the inflow surface and 5 times the hull length from the outflow surface, as shown in Fig. 3.10.

Due to the lack of experimental data for full-scale submarines, dimensionless numbers were used for comparison with experimental and literature data. Two dimensionless numbers were proposed: the pressure coefficient and the friction coefficient, expressed in Eqs. (3.2) and (3.3), respectively. The focus was on the pressure and friction coefficients of the upper half of the submarine's longitudinal midsection, as outlined in red in the second image of Fig. 3.11.

$$C_p = \frac{2(P_0 - P_\infty)}{\rho U^2} \tag{3.16}$$

$$C_f = \frac{2\tau_w}{\rho U^2} \tag{3.17}$$

In the above equations, C_p and C_f represent the pressure coefficient and friction coefficient, respectively; P_0 denotes the static pressure of the flow field; P_∞ represents the far-field pressure; ρ and U denote the flow field density and velocity, respectively; and τ_w represents the fluid shear stress.

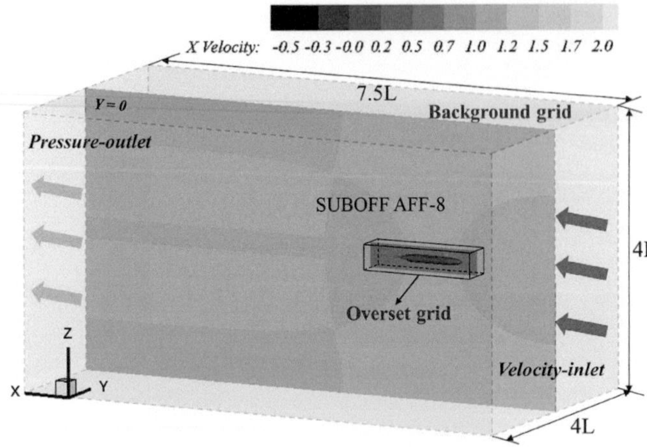

(a) Computational domain setup

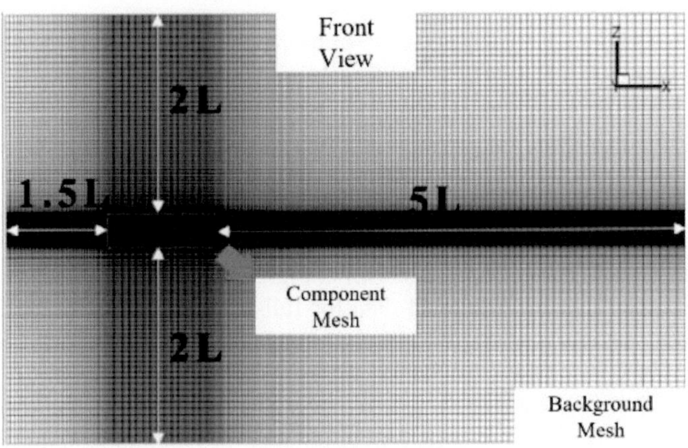

(b) Front view Y=0

Fig. 3.10 Schematic of the overlapping grid independence verification case

The computational results are shown in the curves of Fig. 3.12. It can be observed that the simulation results from the three different grids align well with the experimental data. Compared with simulation results from the literature [115–118], discrepancies exist at the submarine's conning tower and rudder. Analyzing the trend of the numerical solutions and excluding results with significant computational errors (Qiu, 2006), the difference between medium and fine grids was found to be less than 5%. Thus, further refinement beyond 5 million grids has minimal impact on improving accuracy. Therefore, under the premise of ensuring computational efficiency, the medium grid (5 million) is sufficient to meet the computational requirements, and the accuracy of the overlapping grid method is proven to be reliable.

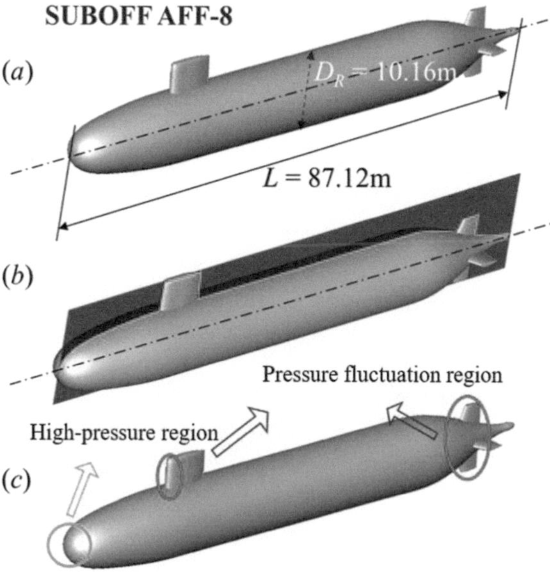

Fig. 3.11 Schematic of the SUBOFF model

Moreover, Fig. 3.12 reveals that the bow of the submarine is a high-pressure zone, while areas near the conning tower and stern rudder exhibit pressure fluctuations (Fig. 3.11c). By examining the dimensionless wall distance y^+ contour of the submarine, we observe that the near the conning tower and stern rudder reaches 300–500, as shown in Fig. 3.13. Based on the turbulence model introduced earlier, the Standard $k-\varepsilon$ turbulence model requires y^+ to be within the range of 30–300. Further optimization of the boundary layer in complex regions of the SUBOFF submarine (e.g., the conning tower and stern rudder attachments) was conducted. The optimized hull's y^+ values fall within 10–20, with a maximum y^+ of 300, meeting the application conditions of the Standard model under high Reynolds numbers.

3.8 Validation of Fluid–Structure Interaction Method

3.8.1 Fluid–Structure Interaction Method

Generally, in studying the interaction between internal solitary waves and submerged bodies, only the motion of the submerged body under the influence of internal solitary waves (fluid) is considered, while the effect of body deformation on the flow field is neglected. This can broadly be regarded as a fluid–structure interaction problem.

In Fluent, such problems are addressed using one-way coupling. One-way fluid–structure coupling is typically employed when the solid deformation's impact on the flow field is negligible. Here, adopting the concept of one-way coupling, the

Fig. 3.12 Variation of dimensionless numbers along the upper half of the submarine's longitudinal midsection

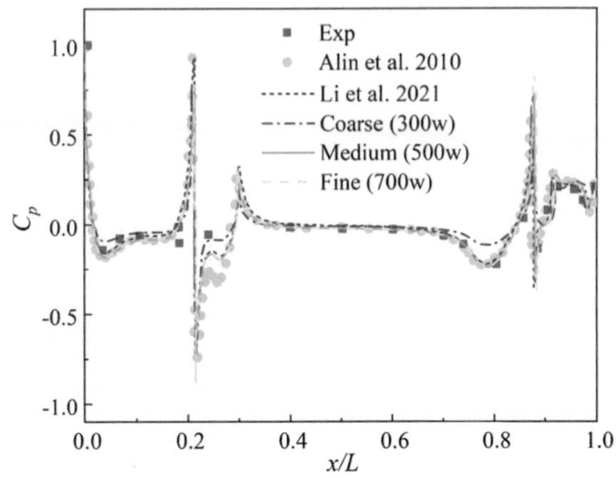

(a) Pressure coefficient

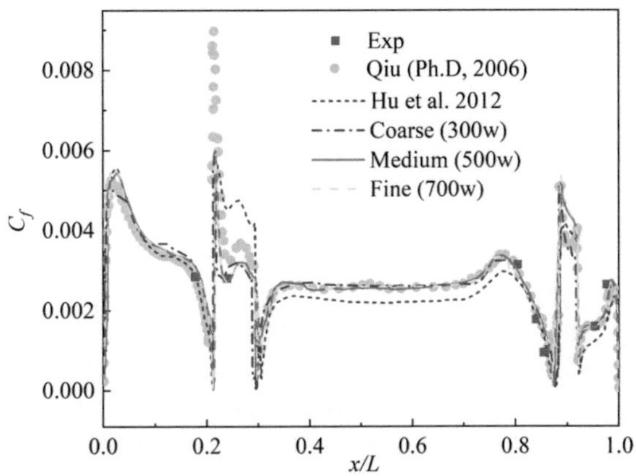

(b) Friction coefficient

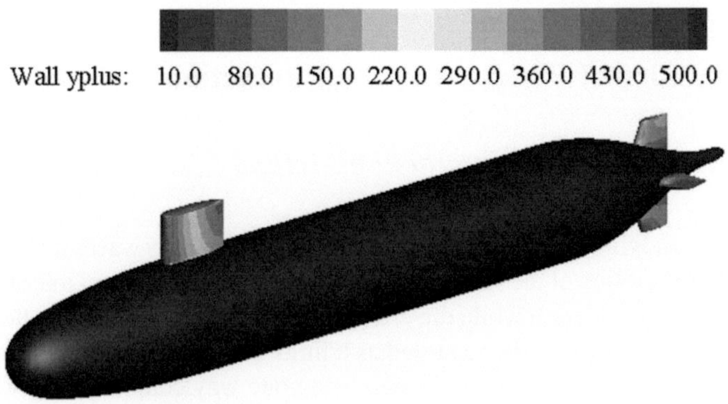

Fig. 3.13 Distribution of submarine wall

submerged body is treated as a rigid body without deformation. Structural deformation parameters are not calculated; only the forces and moments acting on the submerged body are computed to simulate the motion of the rigid body interacting with the fluid.

3.8.2 Case Validation

To verify the accuracy of internal solitary wave loads and the fluid–structure interaction method, a simulation case of a freely moving suspended sphere under internal solitary waves was established based on the experimental work of Cui et al. [103]. The numerical tank dimensions were 10.0 m × 0.4 m × 0.5 m (length × width × height). A sphere with a diameter of $D_s = 8.8\,\text{cm}$ was placed 5.0 cm below the fluid interface, with its center of gravity located $L_s = 5.0\,\text{m}$ from the inlet boundary, as shown in Fig. 3.14. The amplitude of the internal solitary wave was set to 8.3 cm, with an upper-to-lower fluid depth ratio of 2:8—i.e., the upper layer depth $h_1\prime = 0.1\,\text{m}$, the lower layer depth $h_2\prime = 0.4\,\text{m}$, and the densities of the upper and lower layers were $\rho_1\prime = 1030\,\text{kg/m}^3$ and $\rho_2\prime = 1051\,\text{kg/m}^3$, respectively. The dynamic viscosity of the fluid was $\mu\prime = 0.001\,\text{Ns/m}^2$, and the gravitational acceleration was $g = 9.81\,\text{m/s}^2$. The numerical solver used was a transient solver, with the pressure interpolation scheme set to body-force-weighted, the pressure–velocity coupling set to Couple, and the momentum equation discretized using a second-order upwind scheme. The time step was set to 0.005 s.

As can be seen from Fig. 3.15, the vertical distribution of internal solitary wave horizontal velocity and the time-varying curves of horizontal velocity in upper ($z =$

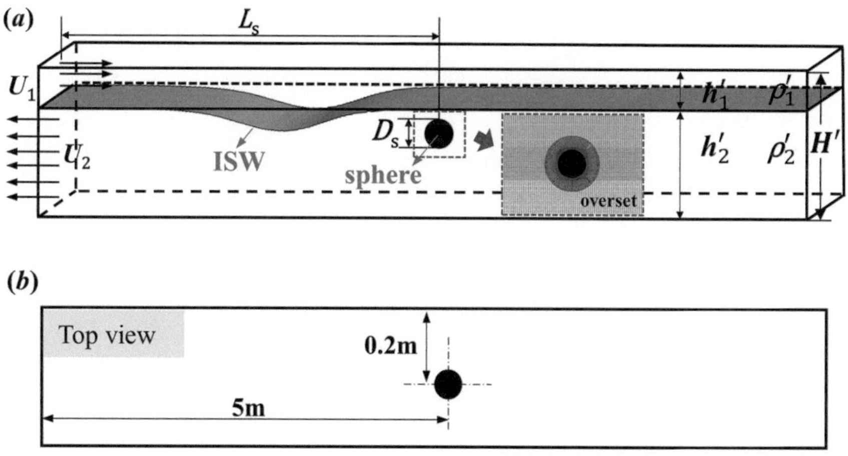

Fig. 3.14 Interaction between internal solitary waves and a suspended sphere **a** Schematic of the case **b** Top view

0.03 m) and lower ($z = -0.2$ m) layers show good agreement with experimental results and theoretical solutions. This indicates that the wave-making method based on MCC equation velocity inlet is accurate and reliable, verifying the accuracy of internal solitary wave flow field and loads, which provides theoretical support for further studying the effects of internal solitary waves with different amplitudes on submerged bodies in the following sections.

In addition, the motion response of a three-degree-of-freedom suspended sphere under internal solitary wave action was simulated using overset grid technology and fluid–structure interaction method. Figure 3.16 reveals that under internal solitary wave action, the suspended sphere first sinks in the negative direction and then rises, with sinking depth and amplitude being similar. The simulation results of surge and heave motions agree well with experimental results, demonstrating that this numerical model is reasonable and feasible for simulating the interaction between internal solitary waves and underwater objects in stratified flows, providing technical support for subsequent studies on internal solitary waves coupled with multi-degree-of-freedom underwater vehicles.

Fig. 3.15 Comparison of simulated horizontal velocity field of internal solitary wave with experimental and theoretical results

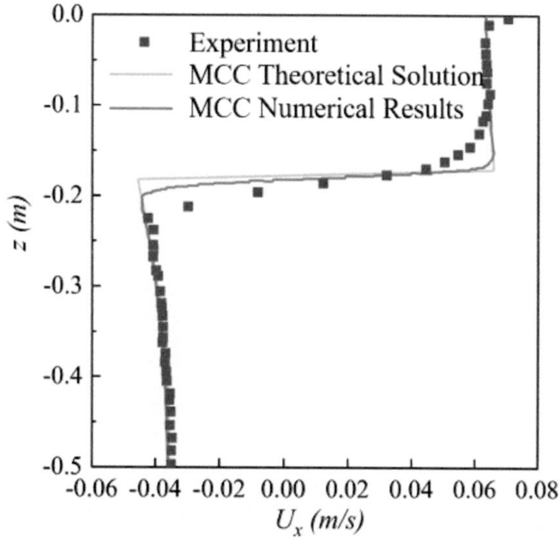

(a)Vertical distribution of horizontal velocity

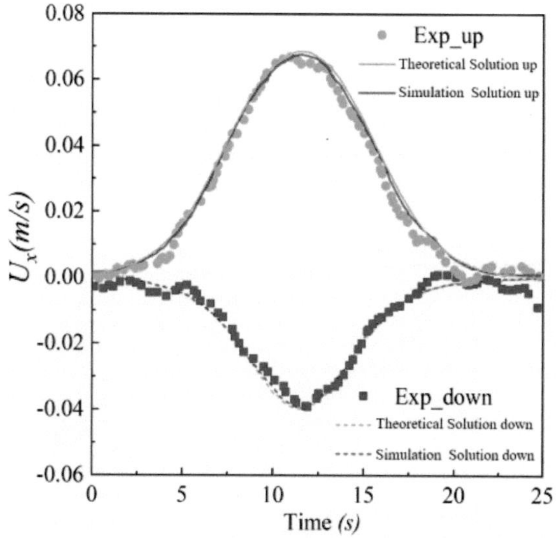

(b)Horizontal velocity in upper and lower layers

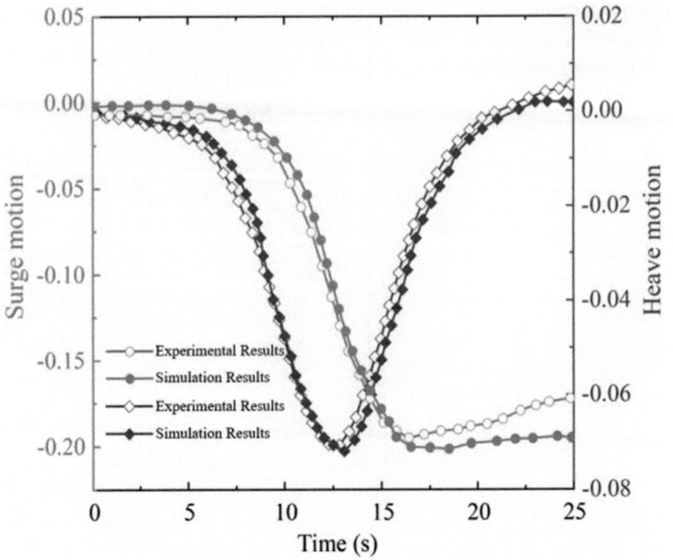

Fig. 3.16 Comparison between simulated and experimental results of sphere's surge and heave motions

Chapter 4
Interaction Between Strongly-Stratified Internal Wave and Submersible

Based on the wave generation method and numerical model proposed above, this chapter mainly studies the interaction between strongly stratified internal solitary waves and suspended/submerged bodies with forward speed. First, the computational model and numerical settings are introduced, and dimensionless transformation is applied to the involved physical parameters. After understanding the model setup, under the principle of controlled variables, the motion and load conditions of suspended and moving submerged bodies at different submergence depths, wave amplitudes, and speeds are studied to analyze the coupling mechanism between strongly stratified internal solitary waves and submerged bodies. Finally, the influence patterns of submergence depth, wave amplitude, and speed on the submerged body are summarized.

4.1 Numerical Model Setup

4.1.1 Computational Model

Based on the verified numerical model and submerged body grid from Chap. 3, a numerical flume for the interaction between strongly stratified internal solitary waves and moving submerged bodies is established. The numerical flume has a length of 5000 m, a width of 200 m, and a height of 300 m (M × N × H = 5000 × 200 × 300 m). It is mainly divided into three regions, from left to right: the wave generation zone, the interaction zone, and the wave absorption zone. The wave generation zone is 3000 m long, the interaction zone is 1000 m long, and the wave absorption zone is 1000 m long, with the width and height consistent with the flume dimensions. The initial position of the submerged body is at the center of the interaction zone, with its center of gravity 3500 m from the inlet wall and 100 m from the front and rear walls, as shown in Fig. 4.1.

P. Du et al., *Internal Wave Hydrodynamics on Submersibles*,
https://doi.org/10.1007/978-981-95-4479-0_4

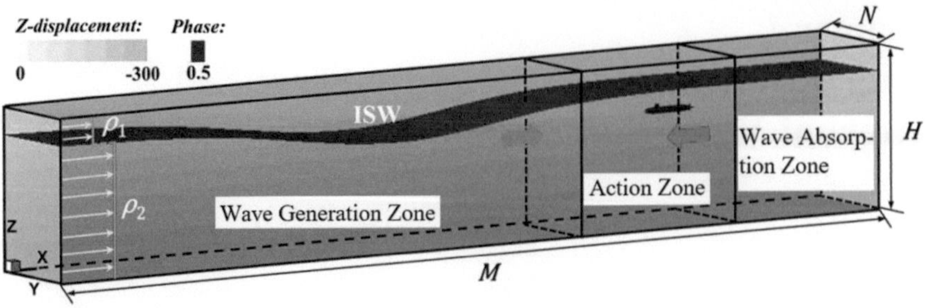

Fig. 4.1 Schematic diagram of the numerical model for the interaction between strongly stratified internal solitary waves and a submerged body

Since this chapter studies the interaction between strongly stratified internal solitary waves and submerged bodies, when simulating internal solitary waves, the thickness of the density interface is neglected, and the upper and lower fluid layers are treated as two immiscible liquids. The upper layer has a thickness of $h_1 = 50\,\mathrm{m}$ and a density of $\rho_1 = 1020\,\mathrm{kg/m^3}$, while the lower layer has a thickness of $h_2 = 250\,\mathrm{m}$ and a density of $\rho_2 = 1028\,\mathrm{kg/m^3}$, with a dynamic viscosity of $\mu = 0.001\,\mathrm{Ns/m^2}$. A three-dimensional Cartesian coordinate system xyz is established, with the vertical plane containing the submerged body as $y = 0$, the tank inlet wall as $y = 0$, and the fluid interface as $z = 0$. The origin of the coordinate system o is shown in Fig. 4.2. The direction of internal solitary wave propagation is defined as the positive x-axis, vertically upward as the positive z-axis, and outward perpendicular to the vertical plane $y = 0$ as the positive y-axis. Counterclockwise rotation of the submerged body is defined as the positive direction of torque. The vertical distance between the submerged body's center of gravity and the interface is defined as the submergence depth d, where $d > 0$ indicates the body is above the interface and $d < 0$ indicates it is below. The distance between the lowest point of the internal solitary wave trough and the interface is defined as the wave amplitude a. Before the internal solitary wave stabilizes, the submerged body remains stationary. The horizontal distance between the submerged body and the central vertical line of the internal solitary wave trough is defined as D_λ. When D_λ equals 1.1 times the characteristic wavelength, the submerged body is precisely located on the vertical line where the internal solitary wave begins to depress, i.e., $D_\lambda = 1.1\lambda = 655$. At this point, the three degrees of freedom of the submerged body are released, and the moment the body begins to move is defined as the time zero. The submerged body with speed maintains a constant horizontal velocity V_x moving leftward.

To simplify subsequent data analysis, eliminate dimensional effects, and quantitatively describe the mechanism of internal solitary wave action on the submerged body, all physical parameters are nondimensionalized here [74, 113]:

$$\text{Variable parameters: } d^* = \frac{d}{h_1}, a^* = \frac{a}{h_1}, V^* = \frac{V}{\sqrt{gL}} \tag{4.1}$$

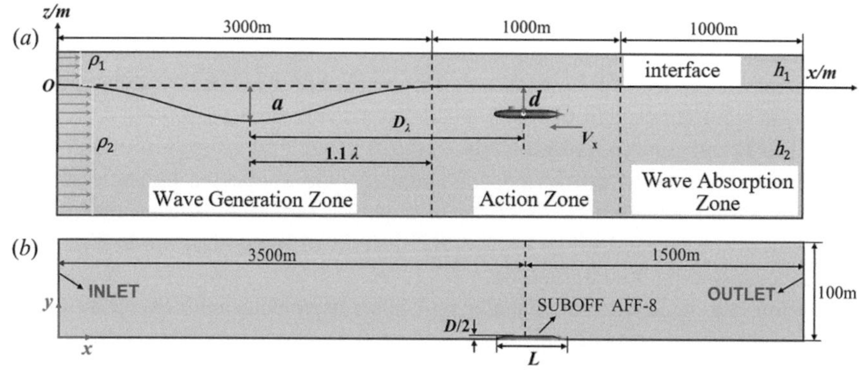

Fig. 4.2 Description of the computational domain **a** Front view of the vertical plane **b** Top view

$$\text{Motion parameters:}\, x^* = \frac{x}{M}, z^* = \frac{z}{H}, T^* = \frac{T}{\lambda/c_i} \tag{4.2}$$

$$\text{Load parameters:}\, F_x^* = \frac{F_x}{\overline{\rho} g S_2 H}, F_z^* = \frac{F_z}{\overline{\rho} g S_2 H}, M_y^* = \frac{M_y}{\overline{\rho} g S_2 HL} \tag{4.3}$$

In the above equations, physical quantities marked with asterisks represent dimensionless quantities. d—submergence depth; a—wave amplitude; V—navigation speed; g—gravitational acceleration; L—submarine length; x—submarine surge; z -submarine heave; M—water tank length; H—water tank height; T—interaction time between internal solitary wave and submarine; λ—characteristic wavelength of internal solitary wave; c_i—phase velocity of internal solitary wave theoretical solution; F_x, F_z, M_y represent the horizontal force, vertical force and pitching moment acting on the submarine during interaction with internal solitary waves; $\overline{\rho}$ is the average density of flow field; S_2—maximum axial windward area of submarine model.

4.1.2 Numerical Settings

Chapter 3 has verified the grid independence of the submarine and the accuracy of overlapping grids. This chapter uses overlapping grid technology to simulate the instantaneous motion process of a submarine encountering internal solitary waves. The optimized structured grid of the submarine is selected, with 5.8 million grids; The water tank grid division is relatively simple. Besides refining the internal solitary wave propagation path and coarsening the wave-absorbing zone grids to increase dissipation, it is also necessary to ensure smooth transition between grid sizes in the interaction zone and submarine boundary grids, which can guarantee higher accuracy during overlapping grid interpolation. After refining the interaction zone, the total number of water tank grids is about 15 million; Due to the large total number of grids and the inevitable need for considerable computational resources for dynamic grid motion, all simulations in this paper are conducted using supercomputing servers for parallel CFD calculations. Additionally, in the dynamic grid settings, the submarine is defined as a rigid body, with its mass and moment of inertia set through user-defined functions. When the submarine's center of gravity is 655 m away from the vertical centerline of the internal solitary wave, the degrees of freedom in three directions (surge, heave and pitch) are released.

The left end of the numerical water tank is set as velocity inlet (Velocity-inlet), and the right end as pressure outlet (Pressure-outlet). Considering the small density difference between upper and lower fluid layers, the vertical velocity induced by internal solitary waves at the free surface during propagation can be basically ignored. Therefore, the top of the fluid domain adopts rigid-lid assumption, with both top and front/rear walls set as symmetry boundaries (Symmetry), and the bottom set as rigid impermeable wall (Wall).

The average horizontal velocities and of upper and lower fluids induced by internal solitary waves during propagation are calculated through MCC theoretical equations. The velocity control equations are implemented through user-defined functions (UDF-User-defined functions) are loaded onto the inlet boundary for numerical wave generation. Initially, the velocity field in the tank is set to zero. A velocity-inlet boundary condition is applied at the inlet, and the pressure-outlet method is used to handle steady inflow and outflow problems. Meanwhile, the sponge layer method is employed for numerical wave absorption of internal solitary waves. After some time of computation, the internal solitary wave forms at the interface between the two fluid layers, necessitating consideration of the two-phase interface tracking problem. The VOF (Volume of Fluid) method is adopted to capture the changes in the two-phase interface.

The numerical solver is configured as an unsteady implicit solver. The governing equations are discretized using the finite volume method, with the explicit volume fraction model selected. Time discretization employs a first-order implicit scheme, and the pressure–velocity coupling is set to "Coupled." Node pressure gradient calculation uses the least squares method, and the pressure interpolation scheme is chosen as "Body Force weighted." The momentum equation, convection term, and turbulence transport term are all set to second-order upwind schemes to control computational accuracy. The time step is set to 0.1 s.

4.2 Simulation of Interaction Between Internal Solitary Waves and Suspended Submerged Bodies

To explore the influence mechanism of internal solitary waves on underwater submerged bodies, this chapter first takes suspended submerged bodies as the research object, preliminarily analyzing the force and motion process of suspended submerged bodies under the influence of internal solitary waves, with a focus on studying the effects of submersion depth and wave amplitude on their motion response and load characteristics. On this basis, further research is conducted on the interaction between moving submerged bodies and internal solitary waves, simulating scenarios where submerged bodies encounter internal solitary waves during navigation, and analyzing the changing patterns of internal solitary waves' influence on the motion of moving submerged bodies. Finally, the load characteristics of submerged bodies moving at different speeds in a straight line through strongly stratified internal solitary waves in the vertical plane are studied, revealing the maneuvering laws for maintaining balance in straight-line motion.

4.2.1 Influence of Submersion Depth

To study the influence of submergence depth on submarine motion and forces, the simulation maintains constant wave amplitude and zero speed. By adjusting the submarine's density (correspondingly changing its mass and moment of inertia), the submarine is fixed at five different submergence depths (Case1–5). When the submarine is in the upper fluid layer, its density is set to 1020 kg/m^3; when in the lower fluid layer, the density is 1028 kg/m^3; when at the interface between two fluid layers, thc density is $\overline{\rho} = 1024\,\text{kg/m}^3$. $d = +25$ m represents the submarine above the interface, $d = 0$ m exactly at the interface, while $d = -25$ m, $d = -50$ m and $d = -75$ m represent positions below the interface. Specific working condition settings are shown in Table 4.1. When the internal solitary wave propagates to the working area of the computational domain ($D_\lambda = 1.1\,\lambda$), the submarine's three degrees of freedom are released to monitor its surge, heave, and force load variations.

Table 4.1 Settings for submerged floating body at different depths

Case	Submergence depth d	Wave amplitude a	Speed V	Pycnocline thickness D_{pyc}
Case 1	+25 m ($d^* = 0.5$)	50 m	0 m/s	0 m
Case 2	0 m ($d^* = 0$)			
Case 3	−25 m ($d^* = -0.5$)			
Case 4	−50 m ($d^* = -1.0$)			
Case 5	−75 m ($d^* = -1.5$)			

When monitoring the submarine's force loads, it was found that the force data oscillates significantly over time, making direct analysis impossible. Here, Eq. (4.4) is used to process the data: applying an average filter by taking the mean of N points before and after each data point.

$$\overline{Y}(x_i) = \frac{1}{2N+1}[y(x_i - N) + y(x_i - N + 1) \cdots + y(x_i) + y(x_i + 1) + \cdots + y(x_i + N)] \quad (4.4)$$

Figure 4.3 shows the dynamic pressure cloud diagram when $T^* = 0.3$ for different suspended submarines initially encountering an internal solitary wave. The background flow field represents the dynamic pressure field, calculated using Eq. (4.5). Dynamic pressure indicates the kinetic energy per unit volume of fluid particles, while the magnitude of fluid inertial forces is typically measured using Eq. (4.6). Both equations contain the term. Therefore, using dynamic pressure for the background flow field not only reflects the kinetic energy of the fluid surrounding the submarine under the influence of internal solitary waves but also indicates the magnitude of inertial forces provided by the surrounding fluid that causes changes in submarine motion.

$$P_d = \frac{1}{2}\rho v^2 \quad (4.5)$$

$$F_I = \frac{\rho v^2}{L_0} \quad (4.6)$$

where: P_d—flow field dynamic pressure; F_I—inertial force; ρ—fluid density; v—average flow velocity; L_0—characteristic length.

(1) Motion response

Figure 4.4 shows the numerical calculation results of the motion responses of suspended bodies at different submergence depths. From the surge variation curves in Fig. 4.4a, it can be observed that the surge amplitudes of suspended bodies located at the interface ($d^* = 0$) and below the interface ($d^* = -0.5, -1.0, -1.5$) are greater than those above the interface ($d^* = 0.5$). The surge variation trends of bodies below the interface are similar, all moving leftward against the incoming direction of the internal solitary wave. In contrast, the suspended body at the interface initially moves a short distance to the right upon encountering the internal solitary wave before quickly reversing to move leftward. The body above the interface also first moves rightward, then shifts leftward, and finally moves rightward again, exhibiting repeated surge variations with relatively small overall amplitudes. The differing surge behaviors between upper and lower suspended bodies are primarily due to the elliptical flow field induced by the propagating internal solitary wave, where the upper-layer flow aligns with the wave propagation direction, while the lower-layer flow opposes it, as illustrated in Fig. 3.3. Within the same calculation period, the surge curves for $d^* = -1.0$ and $d^* = -1.5$ are nearly identical. The surge rate is highest at $d^* = 0$,

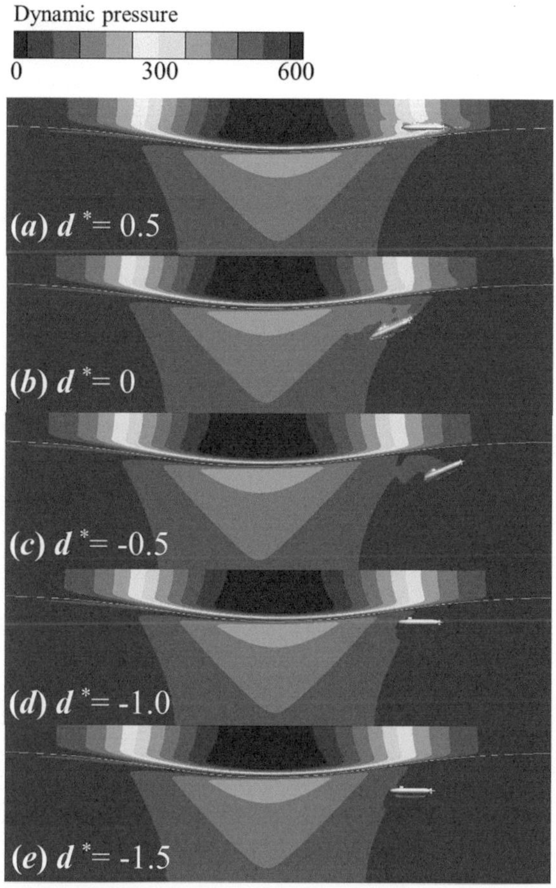

Fig. 4.3 Submarines at different submergence depths initially encountering an internal solitary wave ($T^* = 0.3$)

while at $d^* = 0.5$, it is comparatively gentler, indicating minimal influence from the internal solitary wave.

Figure 4.4b presents the heave variation curves of suspended bodies at different submergence depths. It reveals that under the influence of the internal solitary wave, bodies above the interface first sink and then rise, whereas those below the interface continue to sink. The body at the interface experiences the most significant impact, quickly detaching from the fluid interface and sinking (as shown in Fig. 4.3b). This occurs because the elliptical flow field induced by the internal solitary wave presses the bow of the body downward upon contact, initiating sinking. When the body above the interface reaches the left wave surface, it decelerates and prepares to lift. Within the same duration ($T^* = 0.3$), $d^* = -1.0$ and $d^* = -1.5$ are barely affected by the internal solitary wave, maintaining a horizontal state with nearly identical heave curves. As the internal solitary wave approaches, they begin to slowly lower their bows and sink. After the wave passes, the bodies continue moving downward in

Fig. 4.4 Motion responses of suspended submerged bodies at different depths

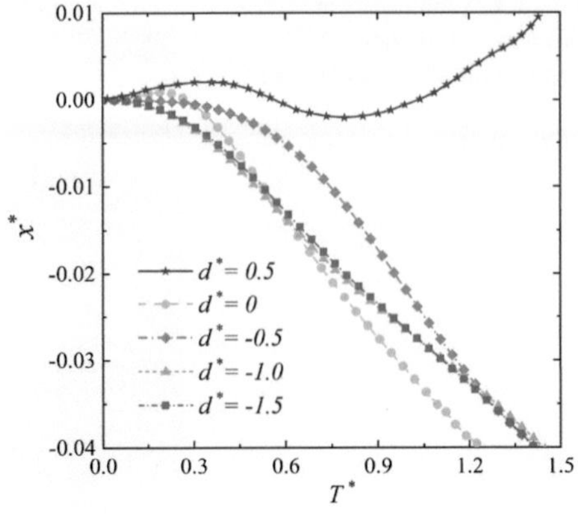

(a) Surge variation

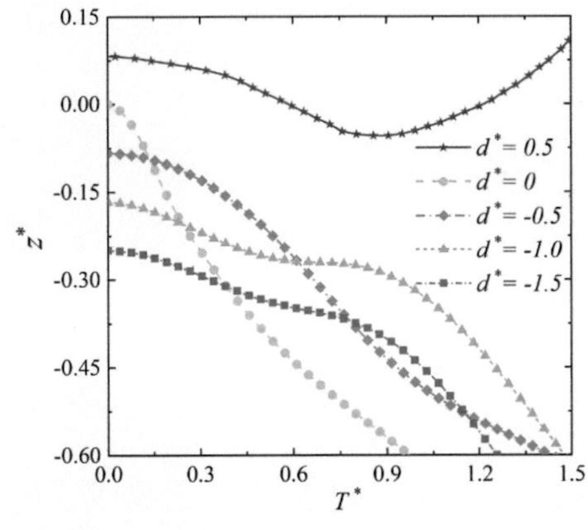

(b) Heave variation

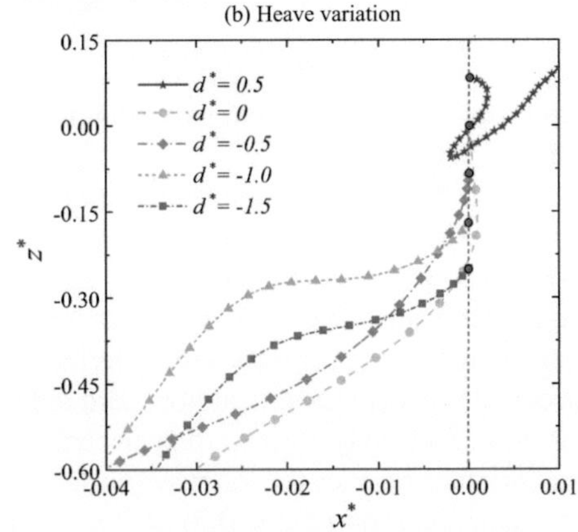

(c) Center of gravity trajectory curve

a lowered-bow state. The heave variation rate is highest at $d^* = 0$, indicating the strongest influence from the internal solitary wave.

Figure 4.4c depicts the trajectory curves of the center of gravity for submerged bodies at five different depths under the action of internal solitary waves, with red dots indicating the initial positions of the bodies. As shown in the figure, the motion trajectories of submerged bodies at depths $d^* = -1.0$ and $d^* = -1.5$ exhibit similar trends—both slowly descend leftward while bowing downward and continue to sink in a bow-down state after the internal solitary wave passes. In contrast, submerged bodies at depths $d^* = 0$ and $d^* = -0.5$ are significantly affected by the internal solitary wave, showing large rates of change in both surge and heave motions. As soon as the internal solitary wave approaches, the submerged bodies begin drifting leftward and downward. The motion trajectory of $d^* = 0.5$ fluctuates around the initial position. When the submerged body descends leftward and contacts the left wave surface, it starts to lift its bow and is then "pushed" rightward by the left wave surface.

(2) Load characteristics

Figure 4.5 shows the numerical calculation results of load characteristics for suspended underwater vehicles at different depths. Figure 4.5a presents the horizontal force variation curves of suspended vehicles at different depths. As analyzed in Sect. 2.4, the horizontal force on the vehicle mainly consists of frictional force, pressure differential force, and fluid inertial force. From the motion response curves of the suspended vehicle in Fig. 4.4, it can be observed that the vehicle's velocity in the internal solitary wave flow field is lower than its velocity in the stratified fluid.

When the vehicle is in the upper fluid layer or at the interface, it first moves to the right. The pressure differential force perpendicular to the vehicle's surface tilts to the left, generating a rightward horizontal force. Additionally, since the vehicle's velocity is lower than the flow velocity in the upper layer of the internal solitary wave field, the surrounding fluid exerts a rightward frictional force. Consequently, the resultant horizontal force acts to the right, causing the vehicle to surge rightward.

As the vehicle approaches the internal solitary wave, the dynamic pressure at its bow becomes greater than at its stern, causing the vehicle to pitch downward. The pressure differential force decreases, and the horizontal force acting on the vehicle begins to diminish. When reaching the left wave surface, the horizontal force approaches zero. At the wave trough center, the fluid velocity increases, and the horizontal force starts to grow leftward.

As the internal solitary wave continues propagating rightward, influenced by the lower fluid layer, the vehicle's bow rises and begins moving upward. At this point, the pressure differential force tilts rightward, generating a leftward horizontal force, while the frictional force remains rightward and dominant. Thus, the total horizontal force acts rightward. When crossing the left wave surface, the vehicle's projected area increases with growing pitch angle, causing the horizontal force to gradually rise to its peak before decreasing after the internal solitary wave passes.

When the vehicle is at the interface, the horizontal force initially increases rightward. However, the vehicle struggles to maintain balance at the interface and quickly

Fig. 4.5 Load characteristics of suspended bodies at different depths

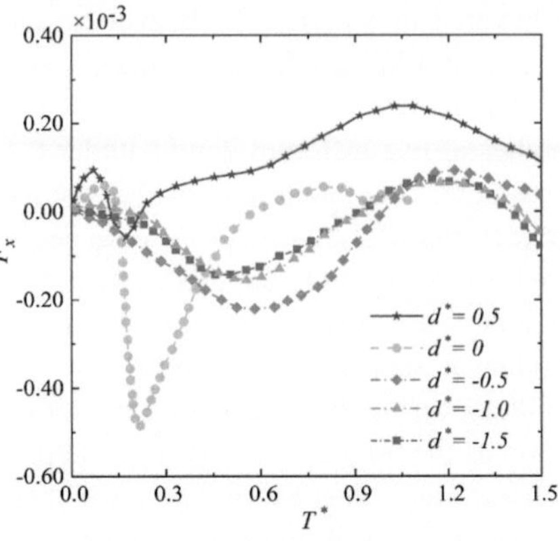

(a) Horizontal force

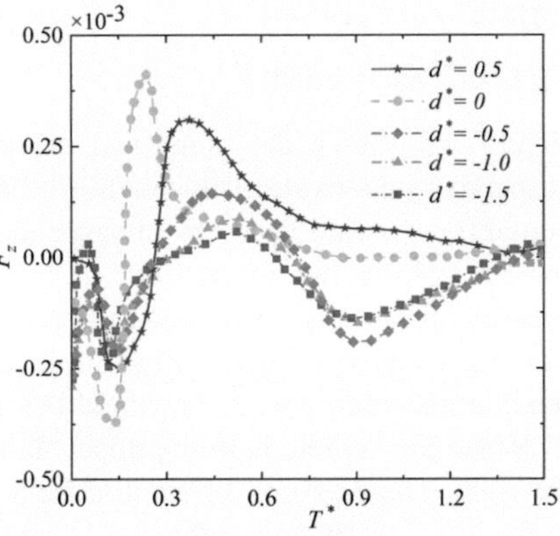

(b) Vertical force

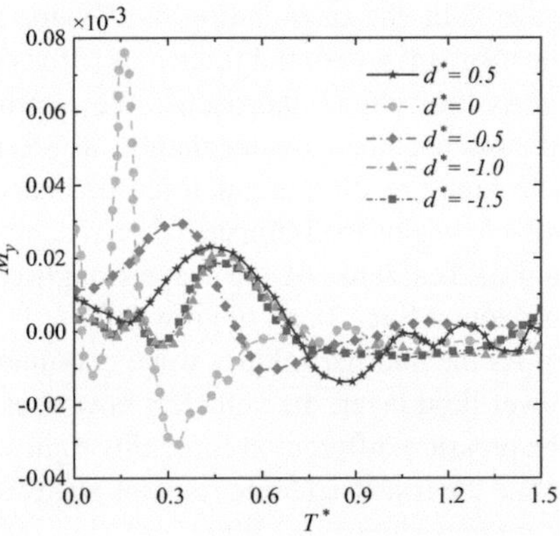

(c) Pitch moment

pitches down, leaving the interface position and entering the lower fluid layer. The lower fluid pushes the vehicle leftward, and the surrounding fluid exerts a leftward opposing frictional force. As the vehicle's pitch angle increases, the leftward horizontal force gradually grows, quickly reaching its peak. Simultaneously, the distance between the vehicle and the internal solitary wave increases, reducing the wave's effect on the vehicle. The horizontal force then begins to decrease and approaches zero. When the vehicle is below the interface, the horizontal force exhibits similar variation trends. Under the influence of the internal solitary wave's lower flow field, the force first increases leftward. After the wave passes, as the vehicle continues pitching down and sinking, its projected area keeps increasing, causing the horizontal force to persistently grow rightward. The difference lies in that $d^* = -0.5$ shows both higher variation rates and peak values of horizontal force compared to $d^* = -1.0$ and $d^* = -1.5$ cases, while $d^* = -1.5$ has the smallest peak. This indicates that the closer the vehicle is to the interface, the greater the influence on its horizontal force.

Figure 4.5b shows the vertical force variation curves of suspended bodies at different submergence depths. Since none of the suspended bodies crossed the internal solitary wave, their buoyancy and gravity remained equal in magnitude. Therefore, the vertical force acting on the submerged bodies was mainly influenced by two factors: first, the inertial force exerted by the disturbed flow field of the internal solitary wave, and second, the pressure difference variation between the upper and lower surfaces of the bodies. From the curves, it can be observed that all bodies initially experienced a downward vertical force. This is because the vertical velocity of the internal solitary wave at the right wave surface was downward, resulting in higher pressure on the upper surface than the lower surface of the bodies. Additionally, the flow field induced by the internal solitary wave was approximately an elliptically rotating clockwise pattern, and the inertial force exerted by the fluid on the bodies was also downward. Thus, as soon as the degrees of freedom were released, the bodies were subjected to a downward vertical force. As the internal solitary wave propagated to the right, the downward vertical force acting on the bodies gradually increased. Subsequently, as the bodies approached the wave trough, the vertical force began to decrease slowly to zero. Among them, the bodies at $d^* = 0.5$, $d^* = 0$, and $d^* = -0.5$ experienced an upward increase in vertical force after passing the wave trough, while the bodies at $d^* = -1.0$ and $d^* = -1.5$ had vertical forces that remained close to zero before increasing downward. This is because the energy of the internal solitary wave was mainly concentrated on the wave surface, and the bodies closer to the wave surface were more significantly affected by the vertical force. When the bodies in the lower fluid reached the left wave surface of the internal solitary wave, they were subjected to an upward fluid force in the vertical direction. However, due to the bodies' own inertia, the actual force direction remained downward. After moving away from the internal solitary wave flow field, the bodies began to sink deeper, with increasing sinking speed, and the vertical force acting on them started to increase downward.

From Fig. 4.5c showing the pitch moment variation curves of suspended bodies at different depths, it can be observed that bodies located at the interface are significantly affected by internal solitary waves, exhibiting large pitch moment variations. After the internal solitary wave passes, the pitch moment shows obvious oscillations due to trailing wave effects. The moment variation trends for $d^* = 0.5$ and $d^* = -0.5$ are similar, both initially experiencing a bow-down moment; at $d^* = 0.5$, when the body contacts the left wave surface, it begins to raise its bow under the influence of lower-layer fluid, while at $d^* = -0.5$, when crossing the wave trough, the upward vertical force from the flow field causes attitude changes—their curves differ by approximately 0.3 T^*. After the wave passes, the bow-up moment gradually approaches zero. The curves for $d^* = -1.0$ and $d^* = -1.5$ are very similar—both initially experience bow-down moments, but due to greater distance from the wave surface, the downward vertical force acts later than other cases, resulting in delayed bow-down. When reaching below the left wave surface, the upward force from the wave is insufficient to raise the bow, so the moment gradually approaches zero.

4.2.2 Influence of Wave Amplitude

To study the effect of internal solitary wave amplitude on body motion and forces, simulations maintained constant depth (20 m below interface) with zero speed. Five simulation cases (Case6–10) with different wave amplitudes were established (see Table 4.2). When waves propagate to the working area ($D_\lambda = 1.1\lambda$), three degrees of freedom are released to monitor surge, heave and load variations.

(1) Motion response

Figure 4.6 shows the submerged body positioned precisely at the trough of internal solitary waves with different amplitudes. From the cloud diagram, it can be observed that the motion posture of the submerged body at the trough position under the action of internal solitary waves with different amplitudes is essentially the same—both involve a downward pitch and sinking. As the wave amplitude increases, the energy of the flow field induced by the internal solitary wave also increases, leading

Table 4.2 Working condition table of suspended submersibles with different wave amplitudes under strong stratified internal solitary waves

Case	Wave amplitude a	Submergence depth d	Speed V	Pycnocline thickness D_{pyc}
Case 6	30 m ($a^* = 0.6$)	−20 m	0 m/s	0 m
Case 7	40 m ($a^* = 0.8$)			
Case 8	50 m ($a^* = 1.0$)			
Case 9	60 m ($a^* = 1.2$)			
Case 10	70 m ($a^* = 1.4$)			

to higher dynamic pressure in both the upper and lower fluid layers, which in turn exerts a greater downward force on the submerged body.

Figure 4.7 presents the numerical calculation results of the motion response of a suspended submerged body under different wave amplitudes. From the surge variation curve in Fig. 4.7a, it can be seen that within the same time frame, the surge amplitude and surge rate are positively correlated with the amplitude of the internal solitary wave, while the surge response time is negatively correlated with the wave amplitude. The larger the amplitude of the internal solitary wave, the greater its range of influence, resulting in an earlier downward pitching force on the submerged body and a faster pitching motion. Figure 4.7b illustrates the heave variation of the submerged body—the larger the wave amplitude, the greater the sinking amplitude and speed. Figure 4.7c depicts the trajectory curves of the submerged body's center of gravity under the influence of five different amplitudes of internal solitary waves, with the red dot indicating the initial position of the submerged body. It is clearly visible that the trajectory curve of $a^* = 1.4$ is more "curved" than that of $a^* = 0.6$,

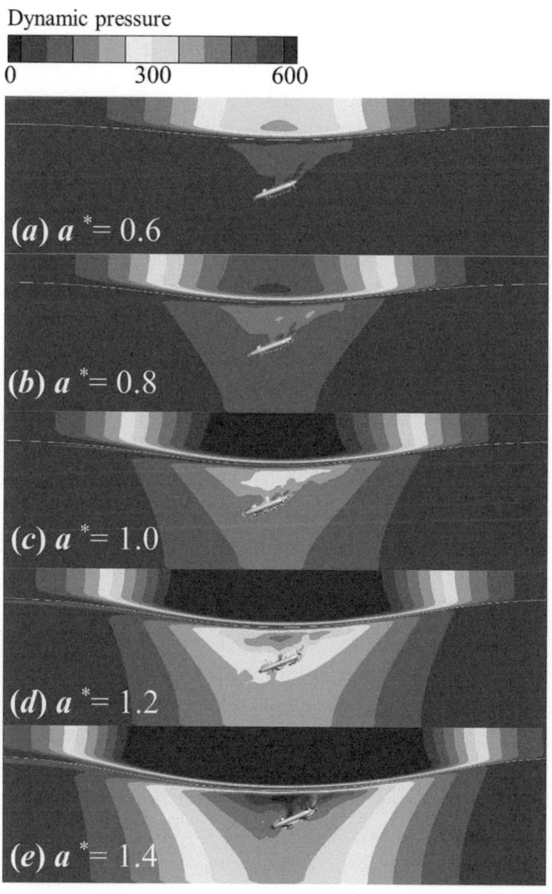

Fig. 4.6 Submerged body located at the trough of internal solitary waves with different amplitudes ($T^* = 0.75$)

indicating that the submerged body is more prone to pitching up under the influence of small-amplitude internal solitary waves. From Fig. 4.6e, it can be observed that the dynamic pressure of the fluid around the bow of the submerged body is relatively high. Under the action of the lower fluid layer, a restoring moment is provided to the submerged body. In cases of smaller wave amplitudes, the restoring moment acting on the submerged body is weaker.

(2) Load characteristics

Figure 4.8 presents the numerical calculation results of the load characteristics of a submerged floating body under internal solitary waves with different amplitudes. Figure 4.8a shows the variation curve of the horizontal force on the floating body. Since all the submerged bodies are located 20 m below the interface, the trends in horizontal force variation are similar. Under the influence of the lower-layer flow field of the internal solitary wave, the horizontal force initially acts to the left. As the submerged body approaches the trough of the internal solitary wave, the central flow velocity increases, causing the horizontal force to rise, primarily due to the viscous friction provided by the surrounding fluid. When the submerged body reaches beneath the left wave surface of the internal solitary wave, the direction of the pressure differential force changes, causing the horizontal force to decrease and approach zero. Under the influence of the lower-layer fluid, the submerged body begins to show a tendency to return to equilibrium. However, the submerged body itself continues to pitch downward, increasing its frontal area, which leads to a rightward increase in the horizontal force. After the internal solitary wave passes, the horizontal force again tends toward zero. The larger the wave amplitude, the greater the peak values of both positive and negative horizontal forces on the submerged body, resulting in more pronounced surge motion in the horizontal direction.

Figure 4.8b illustrates the variation curve of the vertical force on the submerged floating body. Due to the pressure difference between the upper and lower surfaces of the submerged body and the inertial force of the fluid, it initially experiences a downward vertical force. When the submerged body in the lower-layer fluid reaches the left wave surface of the internal solitary wave, it is subjected to an upward vertical force from the fluid. The larger the amplitude of the internal solitary wave, the greater the peak upward vertical force. However, due to the inertia of the submerged body itself, its actual motion remains downward. After moving away from the flow field of the internal solitary wave, the submerged body begins to sink deeper, and the vertical force gradually tends toward zero.

Figure 4.8c displays the variation curve of the pitching moment on the submerged floating body. The trend in moment variation is generally consistent. Initially, the submerged body experiences a bow-down moment. As it approaches the trough of the internal solitary wave, the bow-down moment transitions into a bow-up moment, with the internal solitary wave of $a^* = 1.4$ exerting the most significant bow-up effect on the submerged body. After the internal solitary wave passes, the pitching moment on the submerged body gradually tends toward zero.

Fig. 4.7 Motion response of the suspended submerged body under different wave amplitudes

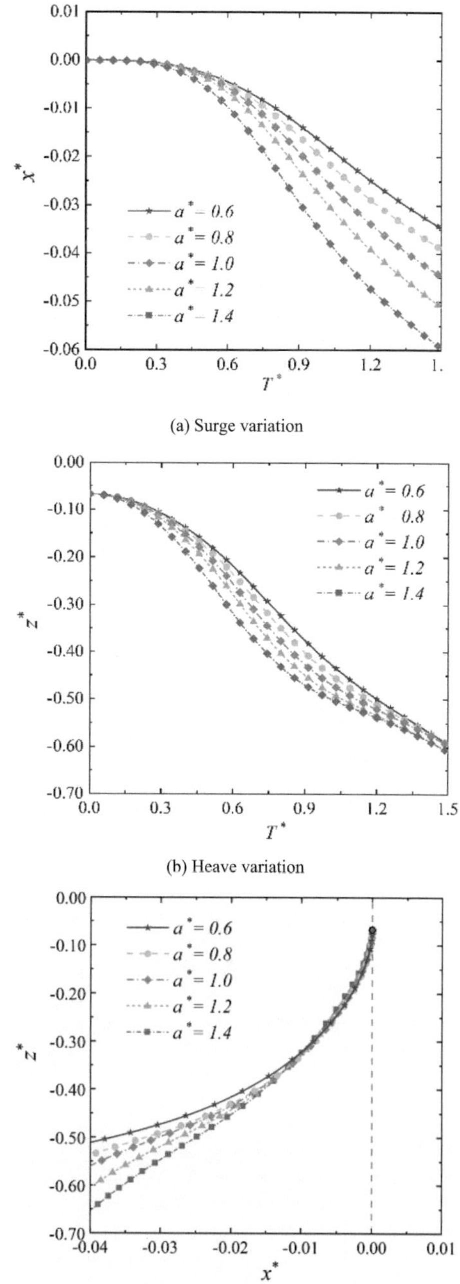

(a) Surge variation

(b) Heave variation

(c) Center of gravity trajectory curve

Fig. 4.8 Load characteristics of the submerged floating body under different wave amplitudes

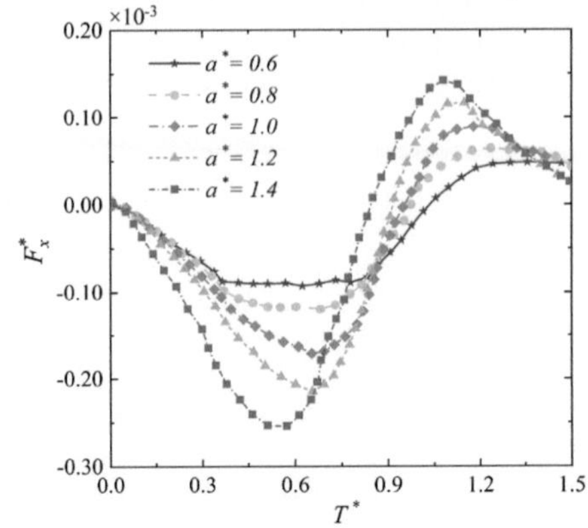

(a) Horizontal force

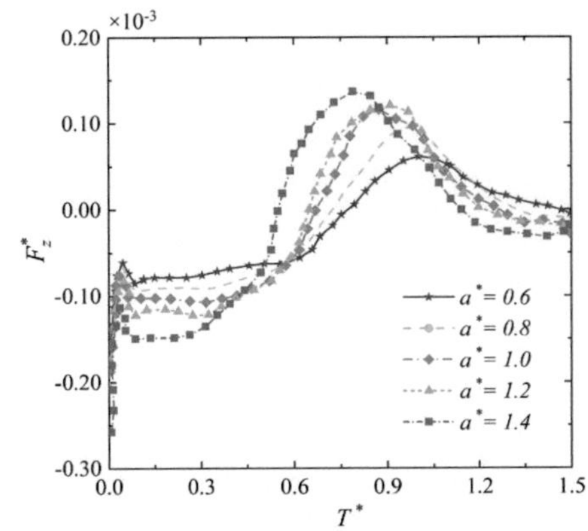

(b) Vertical force

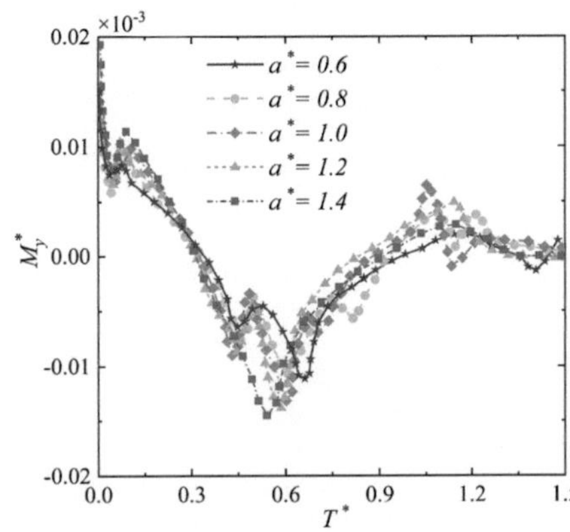

(c) Pitching moment

4.3 Simulation of Interaction Between Internal Solitary Waves and a Submerged Body with Forward Speed

In addition to stationary suspension and hovering operations, underwater vehicles more commonly navigate freely underwater at certain speeds. Generally, moving vehicles encountering internal solitary waves are more prone to dangerous situations, making the study of interactions between freely moving vehicles and internal solitary waves crucial. Building upon the previous simulations of interactions between suspended vehicles and internal solitary waves, this section further investigates the interactions between moving vehicles and internal solitary waves. It simulates scenarios where a vehicle moving at a constant speed encounters internal solitary waves and analyzes the influence of internal solitary waves on the motion and forces experienced by the moving vehicle.

4.3.1 *Influence of Submergence Depth*

To study the impact of submergence depth on the motion and forces of a moving vehicle, the amplitude of the internal solitary wave ($a = 50\,\text{m}$) was kept constant during the simulation. Five simulation cases (Case11–15) with different submergence depths were set up for the moving vehicle interacting with internal solitary waves, as detailed in Table 4.3. When the internal solitary wave propagated into the working area of the computational domain ($D_\lambda = 1.1\lambda$), the vehicle was allowed to move freely in three degrees of freedom at a constant speed of 10 m/s, while its surge, heave, and force variations were monitored.

(1) Motion Response

Figure 4.9 shows the dynamic pressure field cloud diagram of a moving vehicle initially located in the lower layer of the interface interacting with an internal solitary wave. It can be observed that the bow and stern rudder of the moving vehicle experience significant dynamic pressure during navigation. As the vehicle approaches the

Table 4.3 Operating conditions of moving vehicles at different submergence depths under strongly stratified internal solitary waves

Case	Submergence depth d	Wave amplitude a	Speed V	Pycnocline thickness D_{pyc}
Case 11	+25 m ($d^* = 0.5$)	50 m	10 m/s	0 m
Case 12	0 m ($d^* = 0$)			
Case 13	−25 m ($d^* = -0.5$)			
Case 14	−50 m ($d^* = -1.0$)			
Case 15	−75 m ($d^* = -1.5$)			

internal solitary wave, the dynamic pressure on its upper surface becomes greater than that on the lower surface, causing the vehicle to rapidly pitch downward. Due to its speed being much higher than the flow velocity of the internal solitary wave, the vehicle undergoes drifting motion after being disturbed by the wave and quickly sinks to the seabed.

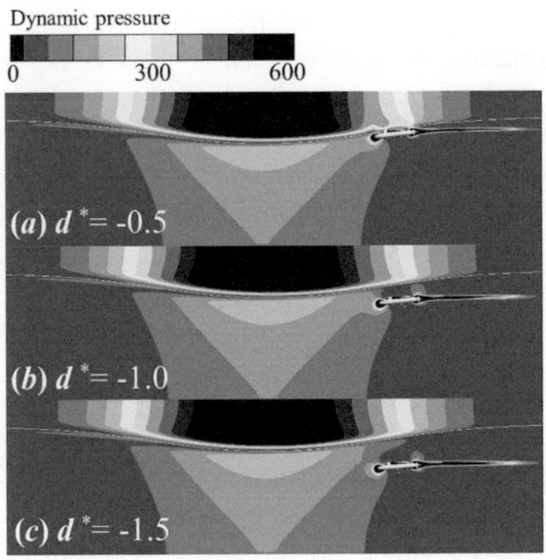

(a) The moving vehicle just contacts the internal solitary wave ($T^* = 0.75$)

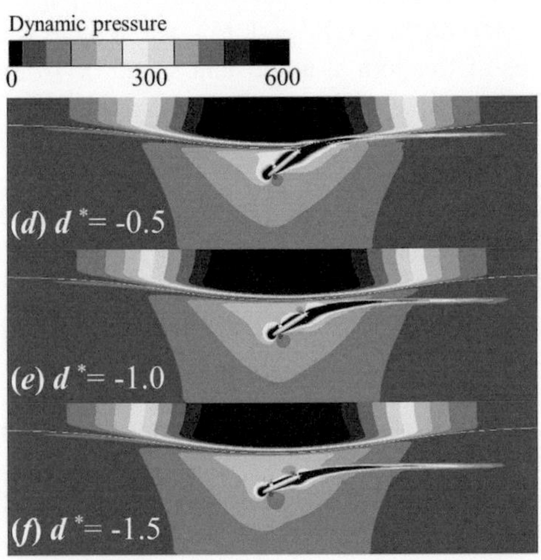

(b) The moving vehicle passes through the trough of the internal solitary wave ($T^* = 0.15$)

Fig. 4.9 Interaction between moving vehicles at different submergence depths and internal solitary waves

Figure 4.10 presents the numerical calculation results of the motion response of moving vehicles at different submergence depths. From the surge variation curve in Fig. 4.10a, it can be seen that the horizontal motion of the vehicle is a straight line because the horizontal speed of the vehicle is set to a constant value (10 m/s). Thus, the horizontal component of the vehicle's motion is uniform linear motion, with the slope of the line representing the horizontal speed of the vehicle.

Figure 4.10b shows the surge curve of the submersible. It can be observed that when the initial depth is below the interface, the submersible sinks downward after passing the trough of the internal solitary wave, as shown in Fig. 4.9II. Within the same time frame ($T^* = 0 \sim 0.25$), the heave amplitude of $d^* = -0.5$ is the largest, followed by $d^* = -1.0$, while $d^* = -1.5$ exhibits the smallest heave amplitude. This is because the farther the submersible is from the interface, the weaker the influence of the internal solitary wave. Compared to a suspended submersible, the towed submersible does not exhibit a slow sinking trend when passing below the trough of the internal solitary wave but instead drifts downward. When the submersible is above the interface ($d^* = 0.5$), it heaves up and down and eventually sinks to the seabed. As the submersible moves against the incoming flow in the upper fluid layer, its center of gravity is closer to the bow. When approaching the right wave surface of the internal solitary wave, the vertical velocity in the flow field is downward. Since the towing speed is much greater than the flow velocity of the internal solitary wave, the effect on the bow is minimal, but the influence on the stern is more pronounced. The dynamic pressure above the stern is relatively high, pushing the submersible to pitch upward. The submersible quickly reaches the area above the trough of the internal solitary wave, where the flow velocity increases. Shortly after leaving the trough, the stern is subjected to an upward force, forcing the submersible to pitch downward. At this point, the submersible loses balance and heaves up and down. When the submersible is at the interface ($d^* = 0$), it initially experiences a small downward pitching moment. As it approaches the right wave surface of the internal solitary wave, the dynamic pressure on the stern increases, causing the submersible to rapidly pitch upward. Upon reaching the trough position, it begins to sink. Figure 4.10c shows the center-of-gravity trajectory curves of submersibles at different depths. It can be seen that when the towed submersible is above the interface, it is significantly affected by the internal solitary wave, exhibiting an oscillatory trend. Meanwhile, the towed submersible at the interface first pitches upward and then rapidly sinks. The motion trajectories of towed submersibles below the interface are similar, with sinking behavior initiated after passing the trough.

(2) Load Characteristics

Figure 4.11 presents the numerical calculation results of the load characteristics of towed submersibles at different depths under the influence of an internal solitary wave. The horizontal force, vertical force, and pitching moment experienced by the towed submersible are two orders of magnitude higher than those of the suspended submersible. Figure 4.11a shows the horizontal force variation curve of the towed submersible. The horizontal force mainly consists of fluid resistance acting to the right. When the submersible's attitude changes, its frontal area increases, leading

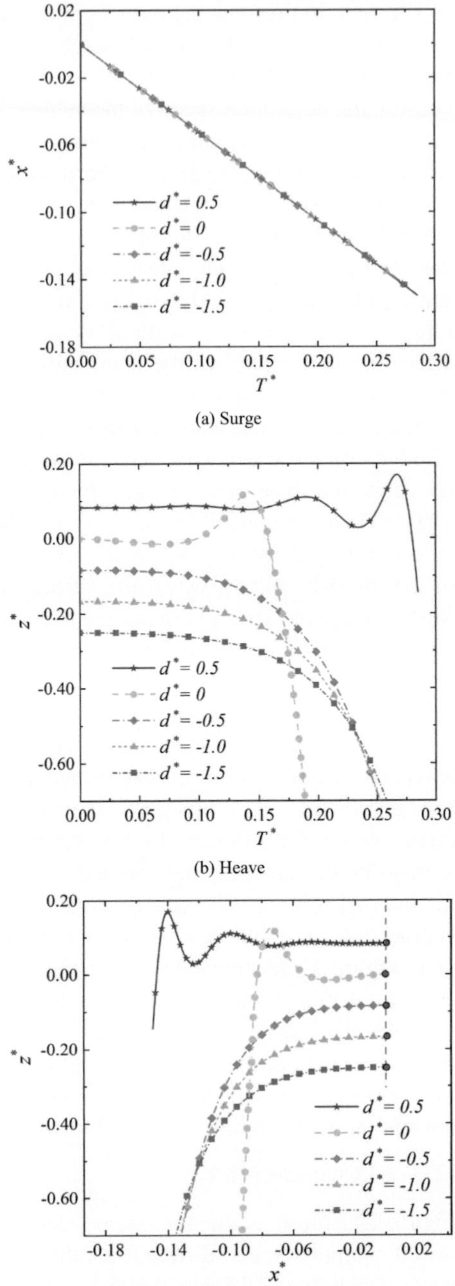

(a) Surge

(b) Heave

(c) Center-of-gravity trajectory curve

Fig. 4.10 Motion response of a towed submersible at different depths to a sharp rise in horizontal force. Once outside the influence range of the internal solitary wave flow field, the horizontal force begins to decrease.

Figure 4.11b shows the vertical force variation curve of submerged bodies with speed. The vertical force variation of submerged bodies above the interface is similar to their horizontal force variation, with larger oscillation amplitudes compared to other depth conditions, ultimately leading to sinking; submerged bodies at the interface initially pitch up, but sink to the trough bottom before reaching the internal solitary wave trough; while submerged bodies below the interface experience continuously increasing downward vertical force, which begins to increase upward after passing the trough, but due to the body's own inertia, they continue to pitch downward.

Figure 4.11c shows the pitching moment variation curve of submerged bodies with speed. The pitching moment variation of submerged bodies is consistent with their vertical force variation, with larger moment oscillations for bodies above the interface; when submerged bodies at the interface sink into the lower fluid layer, under the influence of the lower fluid, their bow-down moment begins to decrease, gradually changing to pitch-up and starting to increase. Submerged bodies with speed below the interface show the same trend, $d^* = -0.5$ the restoring moment acting on submerged bodies at this depth occurs slightly earlier than at $d^* = -1.0$ and $d^* = -1.5$, indicating that when submerged bodies with speed are simultaneously located in the lower fluid layer, the closer they are to the interface, the more pronounced the moment effect they experience.

4.3.2 *Influence of Wave Amplitude*

To study the influence of internal solitary wave amplitude on the motion and forces of submerged bodies with speed, the submerged depth (20 m below the interface) was kept constant during simulation. Five simulation cases (Case16–20) with different internal solitary wave amplitudes interacting with submerged bodies with speed were set up, with specific working conditions shown in Table 4.4. When the internal solitary wave propagates to the working area of the computational domain ($D_\lambda = 1.1\lambda$), the three degrees of freedom of the submerged body are released, moving at a constant speed of 10 m/s, while monitoring its surge, heave and force load variations.

(1) Motion response

Figure 4.12 shows the dynamic pressure field cloud diagrams of submerged bodies with speed interacting with internal solitary waves of different amplitudes. From the figures, it can be observed that,

The amplitude of internal solitary waves significantly affects the motion of submarines with speed; within the same time period ($T^* = 0.15$), under the influence of small-amplitude ($a^* = 0.6$) internal solitary waves, the motion posture of the submarine changes little, whereas under large-amplitude ($a^* = 1.4$) internal solitary

Fig. 4.11 Load characteristics of submerged bodies with speed at different depths

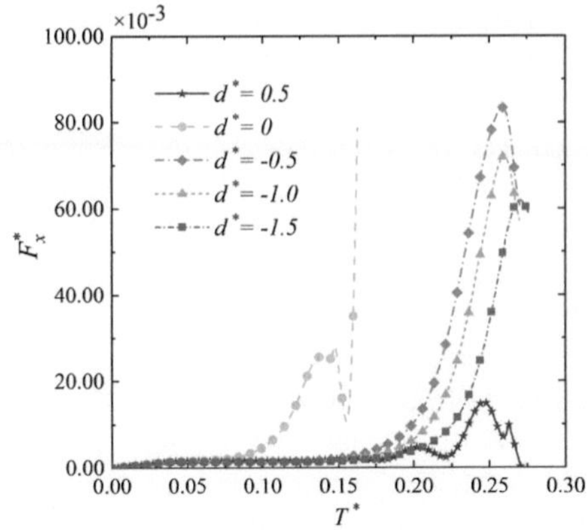

(a) Horizontal Force

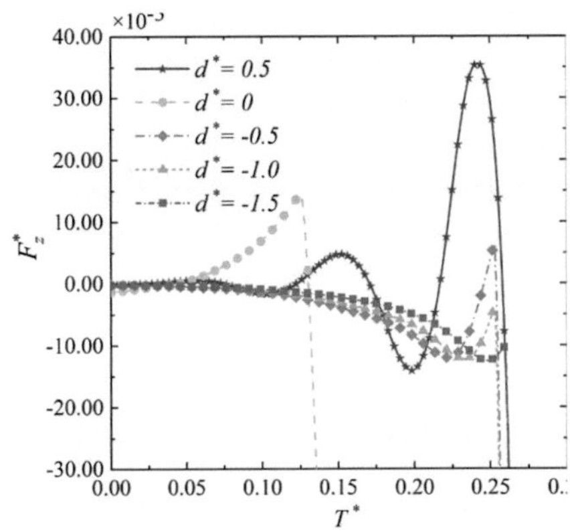

(b) Vertical Force

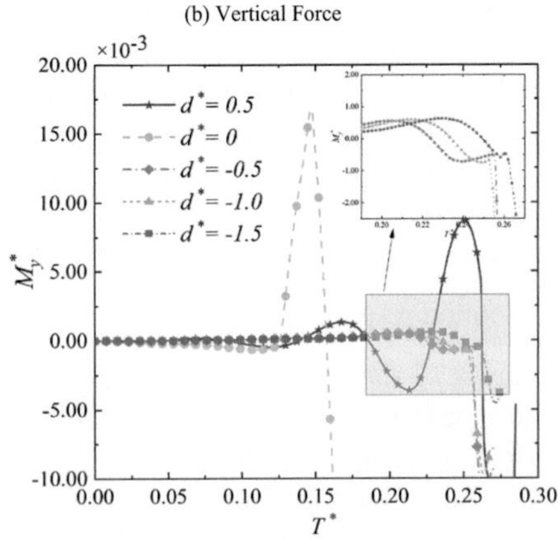

(c)Pitching moment

Table 4.4 Working conditions table for submerged bodies with speed under strongly stratified internal solitary waves with different amplitudes

Case	Wave amplitude a	Submergence depth d	Speed V	Pycnocline thickness D_{pyc}
Case 6	30 m ($a^* = 0.6$)	−20 m	10 m/s	0 m
Case 7	40 m ($a^* = 0.8$)			
Case 8	50 m ($a^* = 1.0$)			
Case 9	60 m ($a^* = 1.2$)			
Case 10	70 m ($a^* = 1.4$)			

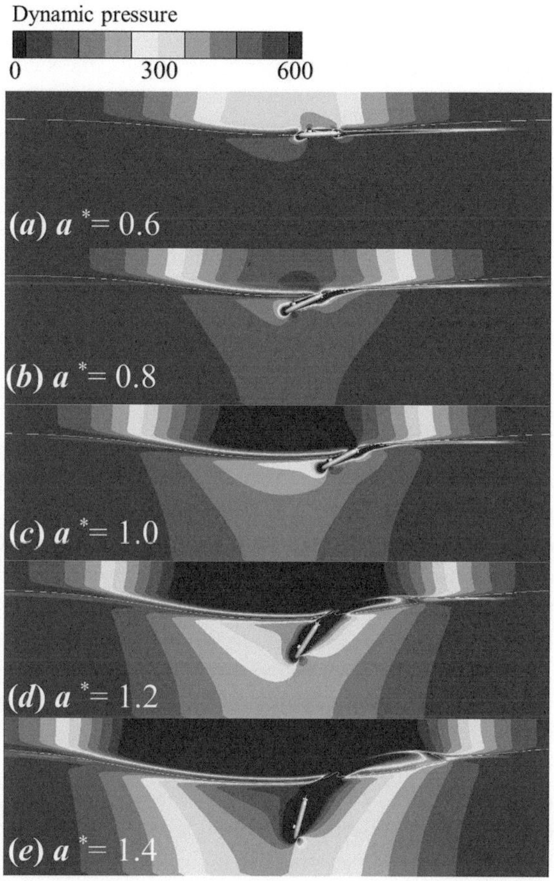

Fig. 4.12 Submerged bodies with speed located at the trough of internal solitary waves with different amplitudes ($T^* = 0.15$)

waves, the submarine's motion posture rapidly undergoes a sudden change, transitioning from horizontal to vertical descent, indicating that large amplitudes pose a substantial threat to the normal navigation of submarines with speed.

Figure 4.13 shows the numerical calculation results of the motion response of a submarine with speed under internal solitary waves of different amplitudes. Since the horizontal speed of the submarine is constant, the surge variation is a straight line (as shown in Fig. 4.13a). The heave variation of the submarine is shown in Fig. 4.13b. Except for the case under small-amplitude ($a^* = 0.6$) waves where the submarine does not pass through the internal solitary wave, in all other amplitude cases, the submarine crosses the right wave surface of the internal solitary wave. When $a^* = 0.6$, the submarine barely grazes the wave surface, showing a tendency to pitch down but quickly corrects its posture and continues navigation with minimal heave variation. In the other four amplitude cases, the submarine pitches down and sinks, with the $a^* = 1.4$ case showing the fastest descent speed and the greatest influence from the internal solitary wave. Figure 4.13c displays the trajectory curves of the submarine's center of gravity under internal solitary waves of different amplitudes. It can be observed that the amplitude greatly affects the motion response of the submarine. Under small-amplitude waves, the submarine's trajectory first pitches down and then up, whereas with increasing amplitude, the submarine continues to sink without control.

(2) Load characteristics

Figure 4.14 presents the numerical calculation results of the load characteristics of a submarine with speed under internal solitary waves of different amplitudes. The load magnitude on the submarine is an order of magnitude higher than that on a suspended submarine. Figure 4.14a shows the horizontal force variation curve of the submarine. It can be seen that the horizontal force is always directed to the right. Under small-amplitude ($a^* = 0.6$) conditions, the horizontal force experiences disturbances after passing through the internal solitary wave. In other larger amplitude cases, as the submarine approaches the internal solitary wave, the horizontal force rapidly increases, with the resistance to the right increasing faster for larger amplitudes.

Figure 4.14b shows the vertical force variation curve of the moving submarine. Under small wave amplitude ($a^* = 0.6$) conditions, the vertical force change is relatively weak when the moving submarine passes through the internal solitary wave. When reaching below the trough of the internal solitary wave, it experiences a small downward force, and after leaving the trough, it receives an upward force due to the action of the lower fluid layer. In other wave amplitude conditions, as the moving submarine approaches the right wave surface of the internal solitary wave, the downward vertical force increases rapidly, causing the submarine to quickly dive downward. When the moving submarine sinks below the trough, under the influence of the internal solitary wave flow field, it experiences an upward vertical force that initially approaches zero and then begins to increase in the opposite direction. As the vertical distance between the submarine and the trough increases, the effect

Fig. 4.13 Motion response of a submarine with speed under different wave amplitudes

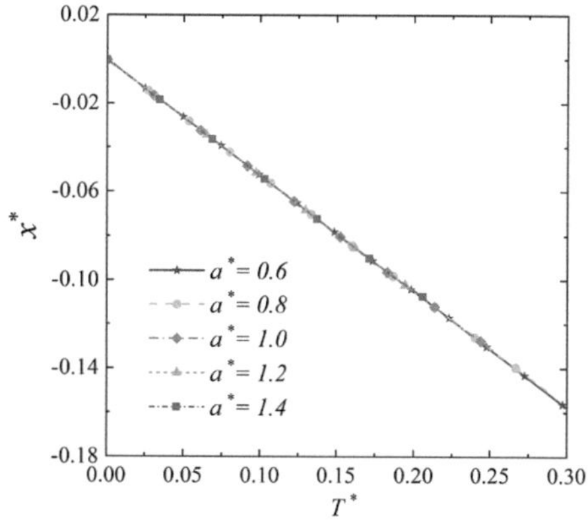

(a) Surge variation

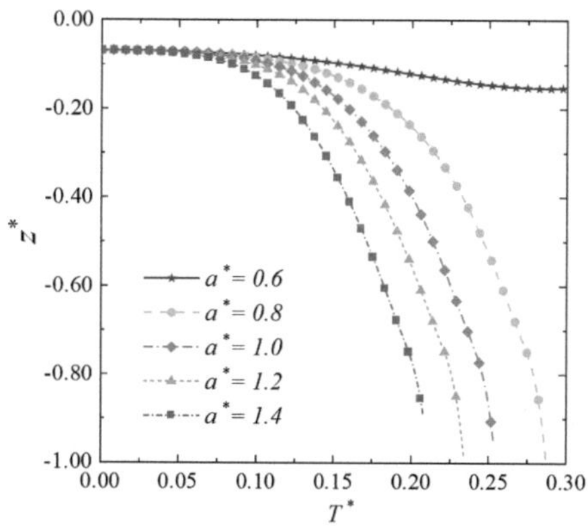

(b) Heave variation

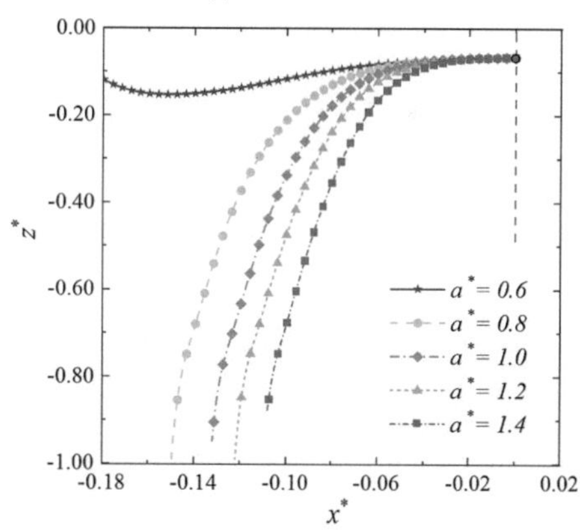

(c) Center of gravity trajectory curve

Fig. 4.14 Load characteristics of moving submarine under different wave amplitudes

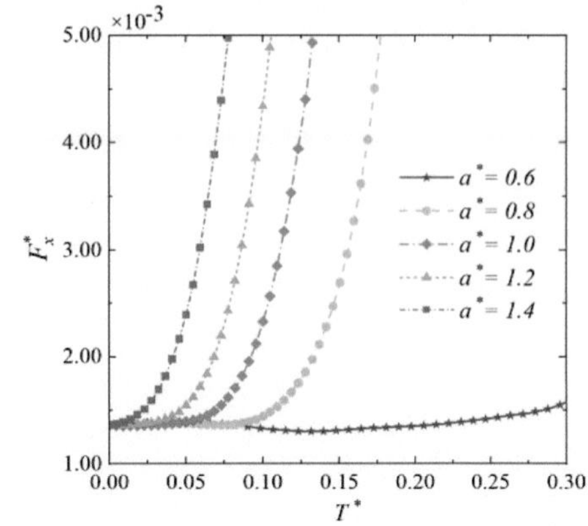

(a) Horizontal force

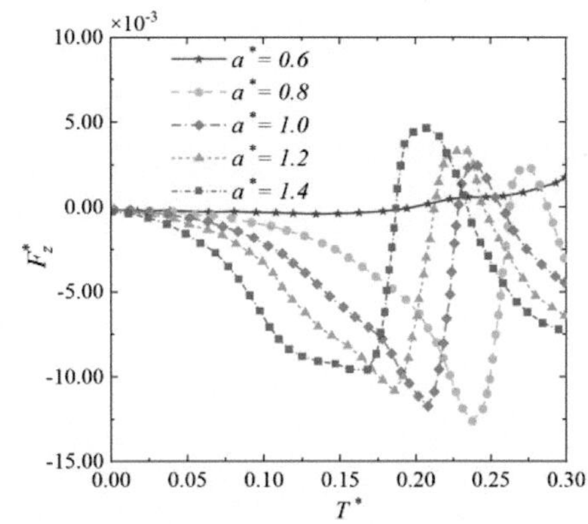

(b) Vertical force

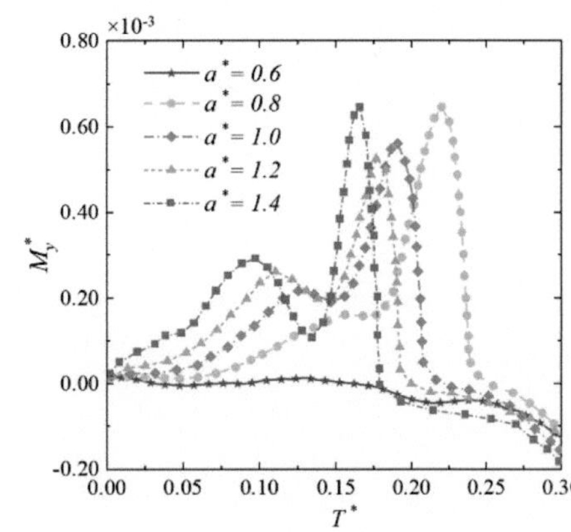

(c) Pitching moment

of the internal solitary wave on the submarine gradually weakens, and the vertical force tends to zero before increasing downward again. For every 0.2 increase in the dimensionless wave amplitude of the internal solitary wave, the influence of the vertical force on the submarine occurs earlier ($0.025T^*$).

Figure 4.14c shows the pitching moment variation curve of the moving submarine. Under small wave amplitude ($a^* = 0.6$) conditions, the pitching moment change of the moving submarine is relatively small. When approaching the right wave surface of the internal solitary wave, the submarine experiences a bow-down moment, which transforms into a bow-up moment after crossing the trough. In other wave amplitude conditions, the submarine continuously experiences a bow-down moment before approaching the internal solitary wave. At the trough position, it begins to sink, and under the influence of the lower fluid layer, the bow-down moment starts to decrease. As the sinking depth increases and the influence of the internal solitary wave flow field weakens, the bow-down moment on the submarine continues to increase. When the submarine's attitude approaches 90°, the pitching moment begins to decrease and tends toward zero.

4.3.3 Influence of Navigation Speed

During most of its underwater movement, such as during cruising, station-keeping, or for facilitating mechanical operations and reducing drag and noise, the submerged body should maintain or approach a state of constant-depth, constant-speed linear motion without trim [127]. Motion with unchanging parameters is called steady motion, which is the focus of this subsection's study on steady motion without trim. Generally, steady motion in the vertical plane can more concentratedly and intuitively reflect the movement and force characteristics of submarines. This section sets the submerged body to traverse internal solitary waves at five different speeds (Case21–25), with specific working conditions listed in Table 4.5. During the movement, constant depth and zero trim are maintained. When the submerged body reaches the same position ($D_\lambda = 1.1\lambda$), its horizontal motion is released while vertical movement and pitch rotation are restricted, allowing monitoring of surface load characteristics.

Figure 4.15 shows the dynamic pressure cloud diagrams of submerged bodies traversing internal solitary waves at different speeds. It can be observed that when the submerged body traverses the waves at higher speeds, the disturbance to the flow field becomes more pronounced, with greater dynamic pressure variations in the fluid above and below the body.

Figure 4.16 presents the load characteristic curves of submerged bodies traversing strongly stratified internal solitary waves at different speeds. From Fig. 4.16a, the horizontal force variation curve reveals that the trend of horizontal force changes is similar across different speeds. When horizontal degrees of freedom are initially released, the submerged body moves linearly at a fixed speed, with horizontal force

Table 4.5 Working conditions for submerged bodies at different speeds

Case	Speed V	Submergence depth d	Wave amplitude a	Pycnocline thickness D_{pyc}
Case 21	5.0 m/s ($V^* = 0.171$)	−25 m	50 m	0 m
Case 22	7.5 m/s ($V^* = 0.256$)			
Case 23	10.0 m/s ($V^* = 0.342$)			
Case 24	12.5 m/s ($V^* = 0.427$)			
Case 25	15.0 m/s ($V^* = 0.513$)			

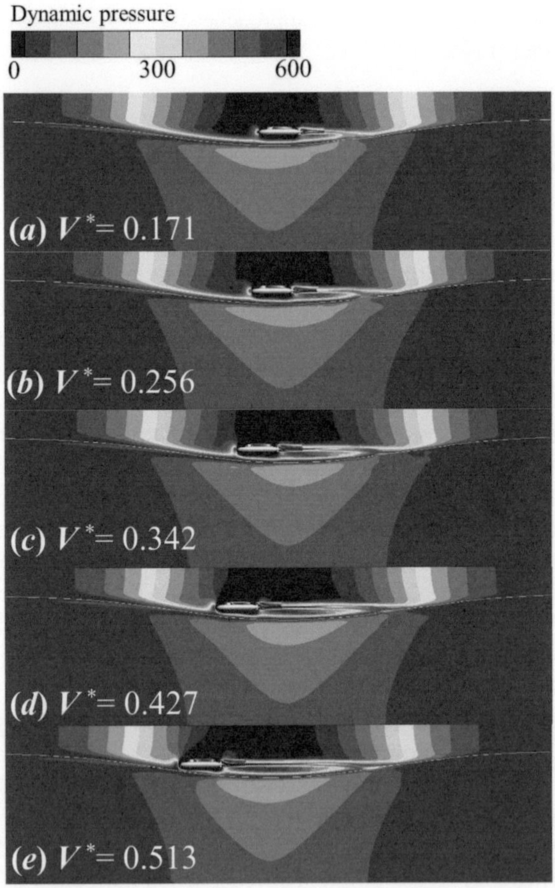

Fig. 4.15 Dynamic pressure cloud diagrams of submerged bodies traversing internal solitary waves at different speeds ($T^* = 0.1$)

gradually increasing as it approaches the internal solitary wave. Upon traversing the right wave surface, as the body moves from lower to upper fluid layers where the flow is rightward, the horizontal force continues to increase. The submerged body moving at speed $V^* = 0.171$ reaches peak horizontal force directly above the wave trough because the flow velocity is fastest there, resulting in maximum friction. With increasing speed, the peak horizontal force grows, and the body's position moves closer to the left wave surface—particularly at speed $V^* = 0.513$, where maximum horizontal force occurs upon impact with the left wave surface. After passing the left wave surface, the horizontal force decreases to a constant value.

Figure 4.16b shows the vertical force variation curve. When a submarine moving at constant speed passes through the internal solitary wave flow field, most of its vertical forces are downward. When crossing the left and right wave surfaces, the vertical force undergoes abrupt changes. When the submarine crosses the right wave surface, transitioning from high-density to low-density water, the buoyancy decreases while the submarine's own gravity remains unchanged, resulting in an increased downward vertical force. When the submarine completely enters the upper low-density fluid layer, the vertical force stabilizes within a certain range. The lower the speed, the longer the submarine remains in the upper fluid layer, maintaining the peak downward vertical force for a longer duration. When crossing the left wave surface, transitioning from low-density to high-density water, the buoyancy begins to increase, causing the vertical force to decrease. As the distance from the internal solitary wave increases, the vertical force gradually approaches zero.

Figure 4.16c displays the pitch moment variation curve. It can be observed that as the speed increases, the peak moment gradually rises, while the duration of the internal solitary wave's effect on the submarine's moment is negatively correlated with speed. Initially, the submarine experiences a bow-down pitch moment. When passing through the trough center of the internal solitary wave, the direction of the pitch moment reverses, causing the submarine to experience a bow-up moment that gradually increases. After moving away from the internal solitary wave, the pitch moment magnitude returns toward zero.

4.4 Summary

This chapter is based on the strongly stratified fully nonlinear internal solitary wave theory—MCC equations, combined with the RANS turbulence model. Using the velocity-inlet wave generation method, a three-dimensional strongly stratified internal solitary wave numerical flume was established to study the effects of three key parameters (submergence depth, wave amplitude, and speed) on submarine motion response and load characteristics. The research shows:

Under the influence of strongly stratified internal solitary waves, submarines located above or at the interface experience significant motion response effects. The closer the submarine is to the interface, the greater the impact on its load characteristics. Wave amplitude primarily affects the speed of submarine motion response

Fig. 4.16 Load characteristics of submarines at different speeds

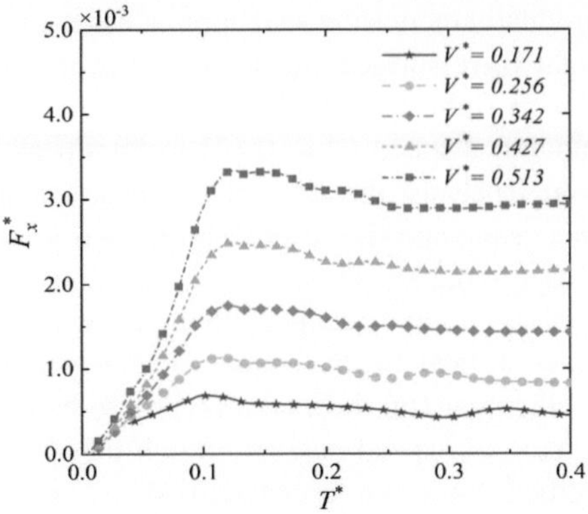

(a) Horizontal force

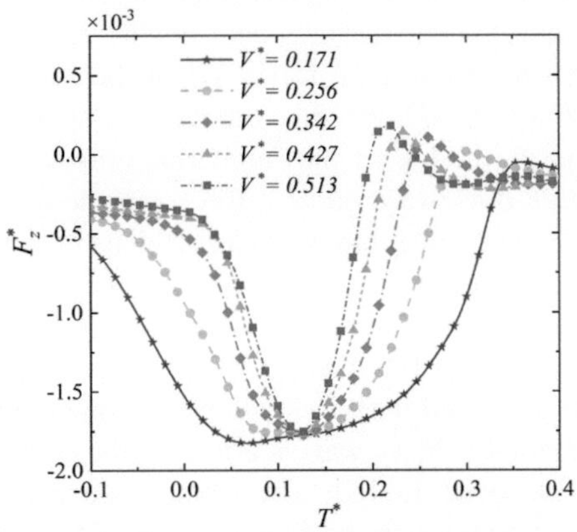

(b) Vertical force

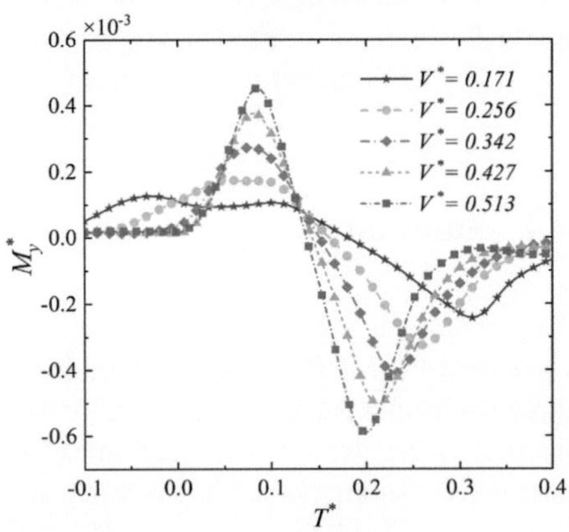

(c) Pitch moment

and the peak load values. Larger internal solitary wave amplitudes result in faster submarine motion responses and higher peak loads. Submarines moving with speed are more prone to loss of control under internal solitary wave influence, experiencing sudden force surges. Higher speeds lead to greater peak loads but shorter wave-body interaction durations.

Chapter 5
Interaction Between Continuously-Stratified Internal Wave and Submersible

In the previous chapter, the thickness of the oceanic pycnocline was neglected to simplify the model, treating the upper and lower seawater as two homogeneous fluid layers. However, in real marine environments, the pycnocline has a finite thickness, and the two-layer model cannot fully characterize the true properties of internal solitary waves. Therefore, this chapter adopts the Dubreil-Jacotin-Long (DJL) equation—a theory for large-amplitude internal solitary waves in continuous stratification—as the wave generation theory. A numerical wave tank for continuously stratified internal solitary waves is established to further investigate the interaction between such waves and suspended or moving submerged bodies. Building upon the analysis in Chap. 4, this chapter summarizes the influence of submergence depth, wave amplitude, and speed on the motion response and load characteristics of submerged bodies under continuously stratified internal solitary waves. It also compares the differences in motion and forces between strongly stratified and continuously stratified conditions, providing valuable references for the safe underwater navigation of submerged bodies in real marine environments.

5.1 Numerical Model Setup

5.1.1 Computational Model

Based on the validated numerical wave tank and submerged body mesh from Chap. 3, a numerical model for the interaction between continuously stratified internal solitary waves and moving submerged bodies is established. The numerical tank has dimensions of 5000 m in length, 200 m in width, and 300 m in height ($M \times N \times H = 5000 \times 200 \times 300$ m), consistent with the tank size and zoning in Chap. 4. The initial position of the submerged body is at the center of the interaction zone, with

P. Du et al., *Internal Wave Hydrodynamics on Submersibles*,
https://doi.org/10.1007/978-981-95-4479-0_5

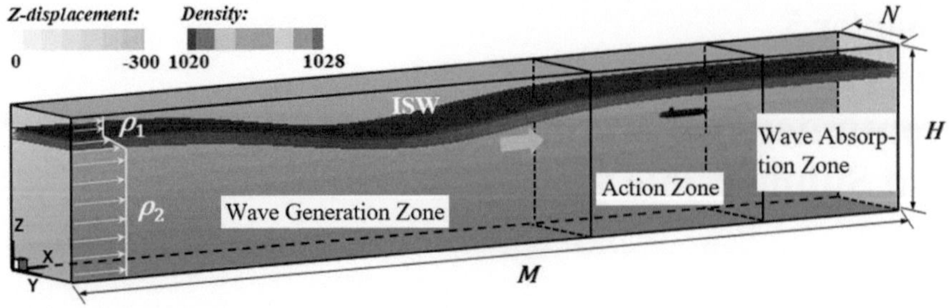

Fig. 5.1 Schematic diagram of the numerical model for the interaction between continuously stratified internal solitary waves and a submerged body

its center of gravity located 3500 m from the inlet wall and 100 m from the front and rear walls, as illustrated in Fig. 5.1.

When simulating continuous internal solitary waves, considering the thickness of the density transition layer, the liquid in the tank can be regarded as three layers. The fluid thickness settings are similar to those of the strongly stratified internal solitary wave tank, with the interface located 50 m below the upper surface of the tank. The wave surface thickness is $D_{pyc} = 20m$, where the density of the upper fluid is set to $\rho_1 = 1020kg/m^3$, and the density of the lower fluid is set to $\rho_2 = 1028kg/m^3$. The density in the transition layer increases progressively from top to bottom, with a density variation range of $1020 \sim 1028kg/m^3$. The dynamic viscosity of the fluid is $\mu = 0.001Ns/m^2$. A three-dimensional Cartesian coordinate system xyz is established with the vertical plane where the submerged body is located as $y = 0$, the tank inlet wall as $x = 0$, and the fluid interface as $z = 0$. The origin of the coordinate system o is shown in Fig. 5.2. Similarly, the direction of internal solitary wave propagation is defined as the positive direction of the x axis, vertically upward as the positive direction of the z axis, and outward perpendicular to the vertical plane $y = 0$ as the positive direction of the y axis. Counterclockwise rotation of the submerged body is defined as the positive direction of the moment. The vertical distance between the center of gravity of the submerged body and the interface (mid-density wave surface of the transition layer) is the submergence depth. Other parameters are the same as those defined in Chap. 4. When D_λ equals 1.1 times the characteristic wavelength, the degrees of freedom in all three directions of the submerged body are released, and the moment when the submerged body begins to move is defined as time zero.

5.1.2 Numerical Settings

The tank and submerged body meshes used in this chapter's simulations are the same as those in Chap. 4. The component meshes of the submerged body are added above the tank mesh, and the submerged body mesh is translated to fix it at the specified

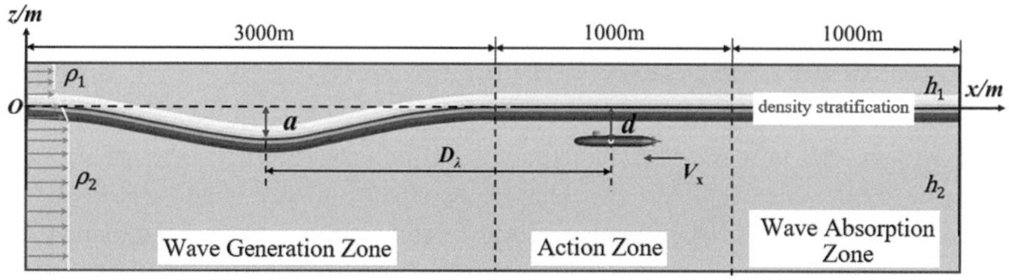

Fig. 5.2 Description of the computational domain

position, creating an overlapping mesh interface. In the dynamic mesh settings, the submerged body is defined as a rigid body, with its mass and moment of inertia set via user-defined functions. When the distance between the center of gravity of the submerged body and the central vertical line of the internal solitary wave reaches 655 m, the degrees of freedom in all three directions (surge, heave, and pitch) of the submerged body are released.

After setting the thicknesses of the upper and lower fluids, the relative density ratio, and the center position of the density transition layer, the desired amplitude of the continuously stratified internal solitary wave is obtained by modifying the available potential energy (APE). The DJL equation is then solved to calculate the initial field, which is interpolated into the Fluent solver to obtain the initial flow field for numerical computation. The bottom of the numerical tank is set as a rigid, impermeable solid wall (Wall), while the remaining walls are set as symmetric boundaries (Symmetry).

Due to the density gradient variation of continuously stratified internal solitary waves at the pycnocline, it can be understood that there are multiple density isosurfaces (wave surfaces) at the pycnocline. Therefore, the mixing and penetration phenomena between different density layers need to be considered. Here, the Mixture multiphase flow model is adopted, which takes into account mass exchange compared to the VOF model. When the submerged body remains stationary, the standard $k-\varepsilon$ turbulence model is used, and the sponge layer wave-absorbing method is added to the source term through a user-defined function (UDF) for numerical wave absorption.

The numerical processing solver is set as an unsteady implicit solver. The governing equations are discretized using the finite volume method, and the implicit volume fraction model is selected. The time discretization adopts the first-order implicit scheme, the pressure–velocity coupling is set to Coupled, the node pressure gradient is calculated using the least squares method, and the pressure interpolation scheme is chosen as Body Force weighted. The momentum equation, convection term, and turbulent transport term are all set to the second-order upwind scheme to control computational accuracy. The computational time step is consistent with that of strongly stratified internal solitary waves, which is set to 0.1 s here.

5.2 Simulation of Interaction Between Internal Solitary Waves and Suspended Submerged Body

To investigate the influence mechanism of continuously stratified internal solitary waves on underwater vehicles, this chapter first takes a suspended vehicle as the research object to preliminarily analyze the force and motion process of the suspended vehicle under the effect of internal solitary waves, with a focus on studying the impact of changes in submergence depth and wave amplitude on its motion response and load characteristics. On this basis, further research is conducted on the interaction between a moving vehicle and internal solitary waves, simulating the scenario where the vehicle encounters internal solitary waves during navigation and analyzing the variation patterns of internal solitary waves' influence on the motion of the moving vehicle. Finally, the load characteristics of vehicles with different speeds moving at a constant speed in a straight line through strongly stratified internal solitary waves in the vertical plane are studied, and the maneuvering rules for maintaining the balance of the vehicle in uniform linear motion are obtained.

5.2.1 *Influence of Submergence Depth*

To study the influence of submergence depth on the motion and forces of the submerged body, the simulation maintains constant wave amplitude and zero speed while varying the density of the submerged body (correspondingly changing its mass and moment of inertia) to fix it at five different depths (Case 1–5). When the body is in the upper fluid layer, its density is set to; when in the lower fluid layer, the density is set to; and when at the interface between the two fluid layers, the density is set to. The specific working conditions are shown in Table 5.1. When the internal solitary wave propagates to the working area of the computational domain (), the three degrees of freedom of the submerged body are released to monitor its surge, heave, and force load variations. The force calculation results are then processed with a mean-value filter.

(1) Motion Response

Table 5.1 Operating conditions for suspended submersibles at different depths

Case	Submergence depth d	Wave amplitude a	Speed V	Pycnocline thickness D_{pyc}
Case 1	$+25$ m ($d^* = 0.5$)	50 m	0m/s	0 m
Case 2	0 m ($d^* = 0$)			
Case 3	-25 m ($d^* = -0.5$)			
Case 4	-50 m ($d^* = -1.0$)			
Case 5	-75 m ($d^* = -1.5$)			

Figure 5.3 shows the numerical calculation results of motion responses for suspended submersibles at different depths. From the surge variation curves in Fig. 5.3a, it can be seen that the surge amplitudes of submersibles above the pycnocline ($d^* = 0.5$) and at the interface ($d^* = 0$) are similar, both smaller than those of submersibles below the pycnocline. The surge motions at depths $d^* = 0.5$ and $d^* = 0$ can be approximately regarded as moving continuously to the right along the propagation direction of the internal solitary wave, with large oscillations but a relatively small overall range. In contrast, submersibles below the pycnocline move leftward against the incoming flow direction of the internal solitary wave. The submersible closest to the interface ($d^* = -0.5$) exhibits the largest surge amplitude and is most intensely affected by the internal solitary wave. The flow field when the suspended submersible approaches the right wave surface of the internal solitary wave is shown in Fig. 5.4, revealing that the submersible quickly transitions from a horizontal motion posture to a bow-down state. Additionally, comparing the surge of submersibles under strong stratification internal solitary waves, it can be found that the continuous stratification results in smaller surge amplitudes than those under strong stratification internal solitary waves.

Figure 5.3b shows the heave variation curves for suspended submersibles at different depths. Under the influence of the internal solitary wave, submersibles above the pycnocline and at the pycnocline initially sink and then begin to rise upon contacting the left wave surface, while submersibles below the pycnocline continue to sink. This behavior is similar to that observed under strong stratification, with the difference being that in continuous stratification, the submersible at the pycnocline ($d^* = 0$) does not detach and sink from the wave surface but instead moves along the waveform, first sinking and then rising. Moreover, the heave rates of submersibles at various depths under continuous stratification are slower.

Figure 5.3c depicts the trajectories of the center of gravity for submersibles at five different depths under the influence of continuous stratification internal solitary waves, with red dots indicating the initial positions of the submersibles. It can be observed that the submersible at depth $d^* = 0.5$ slowly sinks to the right. Upon contacting the left wave surface, it begins to lift its bow and is then "pushed" by the left wave surface to continue moving rightward. The submersible at the median density moves along the waveform, first sinking and then rising, eventually returning to its initial position. In contrast, the trajectories of submersibles below the pycnocline are similar, with the submersibles bowing down and slowly sinking to the left. After the internal solitary wave passes, they maintain a bow-down state and continue to sink.

(2) Load characteristics

Figure 5.5 shows numerical calculation results of load characteristics for suspended submersibles at different depths. When approaching the left wave surface of internal solitary wave, the pressure difference force perpendicular to the submersible's surface tilts leftward in the upper fluid layer and density interface layer, generating a rightward horizontal force on the submersible. Additionally, as the submersible's rightward velocity is lower than the upper layer flow velocity of internal solitary wave, the

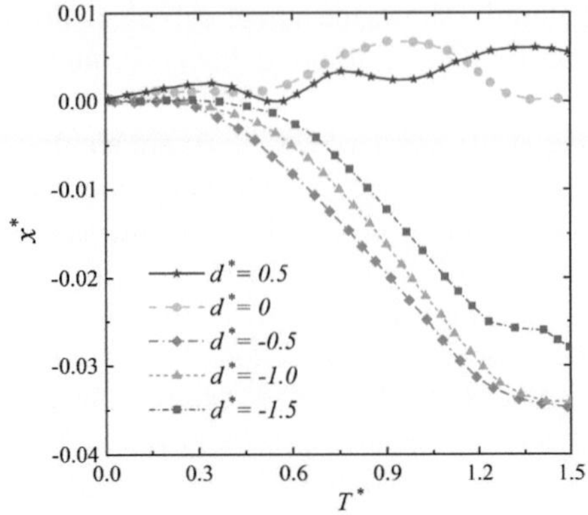

(a)Surge variation

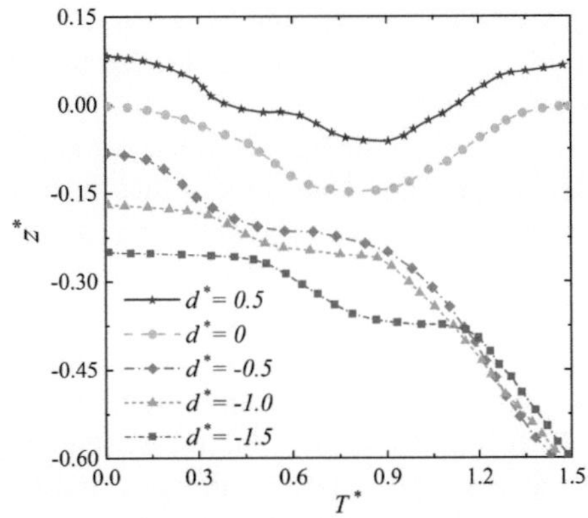

(b)Heave variation

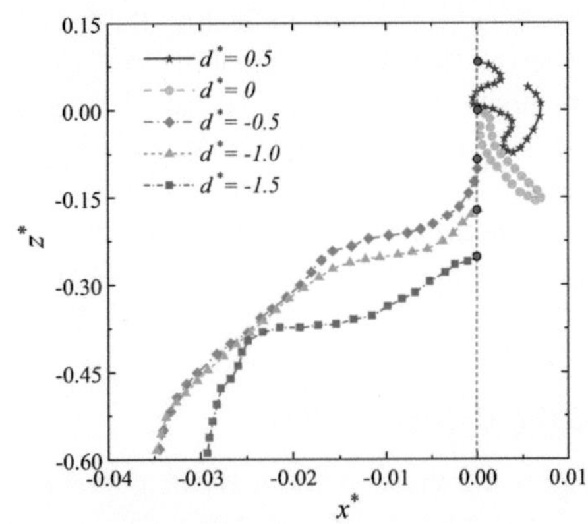

(c)Trajectory curve of center of gravity

Fig. 5.3 Motion responses of suspended submersible at different diving depths

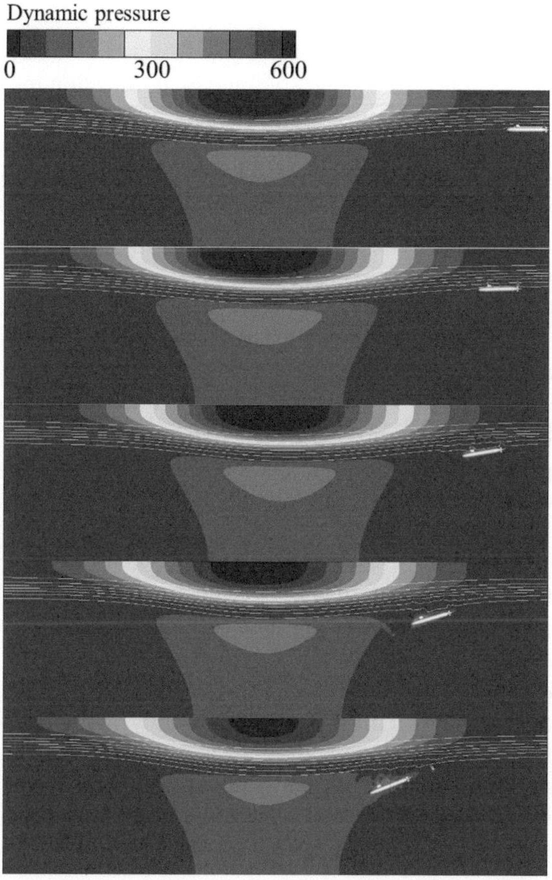

Fig. 5.4 Submersible begins to pitch down when contacting the right wave surface of internal solitary wave ($d^* = -0.5$)

surrounding fluid exerts a rightward friction force, causing rightward surge motion. During the approach to internal solitary wave, the dynamic pressure at bow becomes greater than stern, resulting in bow-down attitude and reduced pressure difference force, leading to decreasing horizontal force. At the left wave surface, horizontal force approaches zero. As fluid velocity increases at wave trough center, horizontal force begins increasing leftward. With continued rightward propagation of internal solitary wave, influenced by lower fluid layer, the submersible's bow rises and begins upward motion, with pressure difference force tilting rightward to generate leftward horizontal force while friction remains rightward (dominant), resulting in net rightward horizontal force. After passing left wave surface, the flow-facing area increases with growing pitch angle, causing horizontal force to gradually reach peak value. For submersibles below density interface layer, horizontal force variation follows similar trend—first increasing leftward under lower layer flow field, then continuously increasing rightward after internal solitary wave passes due to increasing flow-facing area during bow-down sinking. Notably, $d^* = -0.5$ shows greater rate of change and peak value in horizontal force than $d^* = -0.5$ and $d^* = -1.5$, indicating that closer proximity to density interface layer leads to greater influence on horizontal force, and continuous stratification results in smaller horizontal force compared to strong stratification cases.

Figure 5.5b shows the vertical force variation curves of suspended submersibles at different depths. Consistent with the trend under strongly stratified internal solitary waves, the curves indicate that the submersible initially experiences a downward vertical force upon releasing its degrees of freedom. As the internal solitary wave propagates to the right, the downward vertical force gradually increases, then begins to decrease toward zero as the submersible approaches the wave trough. For submersibles at depths $d^* = 0.5$ and $d^* = 0$, after crossing the trough and encountering the left wave face, the vertical force starts increasing upward, causing the submersible to pitch up. In contrast, submersibles below the pycnocline, influenced by the lower-layer flow, first experience an increasing downward vertical force. After passing the internal solitary wave, they show a tendency to move upward but ultimately descend further. The vertical force on submersibles near the pycnocline is significantly affected.

From Fig. 5.5c, which depicts the pitch moment variation curves of suspended submersibles at different depths, it can be observed that submersibles located at or above the pycnocline are greatly affected by the internal solitary wave, exhibiting substantial pitch moment fluctuations. The moment trends for depths d^*=0.5 and $d^* = 0$ are similar—both initially experience a bow-down moment, followed by a bow-up moment after passing the internal solitary wave. The moment curves for submersibles below the pycnocline are very similar: they first experience a bow-down moment, then the moment shifts upward when beneath the trough of the internal solitary wave. After moving away from the wave, the submersible continues to pitch down and sink, with the moment gradually approaching zero.

5.2.2 *Influence of Wave Amplitude*

To study the effect of wave amplitude in continuously stratified internal solitary waves on submersible motion and forces, simulations were conducted with a fixed submersible depth (20 m below the pycnocline), zero speed, and a pycnocline thickness of 20 m. Five simulation cases (Case 6–10) with different wave amplitudes were set up for interaction between internal solitary waves and suspended submersibles, as detailed in Table 5.2. When the internal solitary wave propagates into the working zone of the computational domain ($D_\lambda = 1.1\lambda$), the submersible's three degrees of freedom are released, and its surge, heave, and force variations are monitored.

(1) Motion Response

Figure 5.6 shows the numerical calculation results of the motion response of the suspended submerged body under different wave amplitudes. It can be observed that under the action of continuously stratified internal solitary waves, the surge of the

Fig. 5.5 Load characteristics of suspended submersibles at different depths

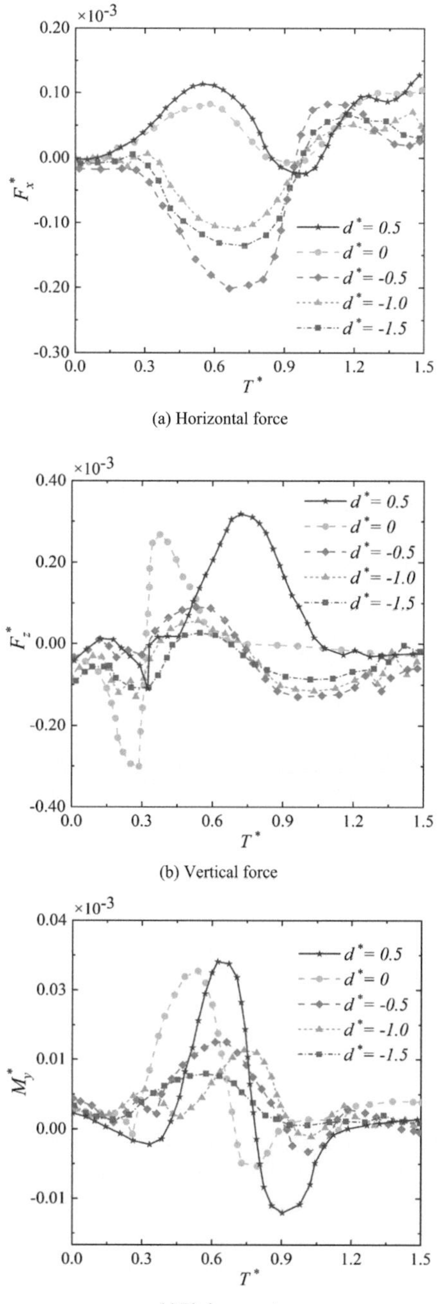

(a) Horizontal force

(b) Vertical force

(c) Pitch moment

Table 5.2 Simulation conditions for suspended submersibles under continuously stratified internal solitary waves with different amplitudes

Case	Wave amplitude a	Submergence depth d	Speed V	Pycnocline thickness D_{pyc}
Case 6	30 m ($a^* = 0.6$)	−20 m	0m/s	20 m
Case 7	40 m ($a^* = 0.8$)			
Case 8	50 m ($a^* = 1.0$)			
Case 9	60 m ($a^* = 1.2$)			
Case 10	70 m ($a^* = 1.4$)			

submerged body slows down from the surge variation curve in Fig. 5.6a. The larger the amplitude of the internal solitary wave, the greater the surge amplitude and the faster the surge response. Figure 5.6b shows the heave variation of the submerged body. Since the submerged body is located below the density interface, it sinks under the influence of internal solitary waves with different amplitudes. As the wave amplitude increases, the magnitude and rate of sinking also increase. Figure 5.6c depicts the trajectory curves of the submerged body's center of gravity under five different internal solitary wave amplitudes. The red dot indicates the initial position of the submerged body. It can be seen that the larger the wave amplitude, the greater the impact on the submerged body's motion.

(2) Load Characteristics

Figure 5.7 presents the numerical calculation results of the load characteristics of the suspended submerged body under different internal solitary wave amplitudes. Figure 5.7a shows the horizontal force variation curve of the suspended submerged body. The trends of horizontal force variation are similar: under the influence of the lower-layer fluid of the internal solitary wave, the horizontal force points to the left. As the submerged body approaches the trough of the internal solitary wave, the central flow velocity increases, causing the horizontal force on the submerged body to rise. At this stage, the horizontal force is mainly provided by the viscous friction from the surrounding fluid. When the submerged body reaches below the left wave surface of the internal solitary wave, the direction of the pressure difference force changes, causing the horizontal force to decrease and approach zero. Under the influence of the lower-layer fluid, the submerged body begins to exhibit a tendency to return to its original position. However, the submerged body continues to sink with a bow-down attitude, and the flow-facing surface area increases, leading to a rightward increase in the horizontal force. After the internal solitary wave passes, the horizontal force tends toward zero again. The larger the wave amplitude, the greater the peak values of both positive and negative horizontal forces on the submerged body, resulting in more pronounced surge variations in the horizontal direction. Compared to strong stratification, the oscillation in horizontal force variation is more noticeable.

Figure 5.7b displays the vertical force variation curve of the suspended submerged body. Due to the pressure difference between the upper and lower surfaces of the

Fig. 5.6 Motion response of the suspended submerged body under different wave amplitudes

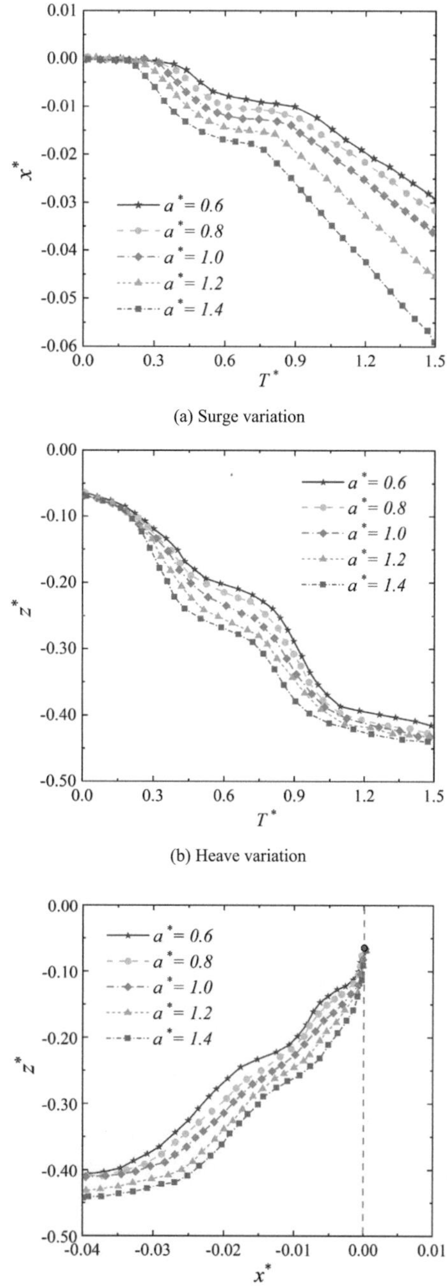

(a) Surge variation

(b) Heave variation

(c) Trajectory curve of the center of gravity

Fig. 5.7 Load characteristics of the suspended submerged body under different wave amplitudes

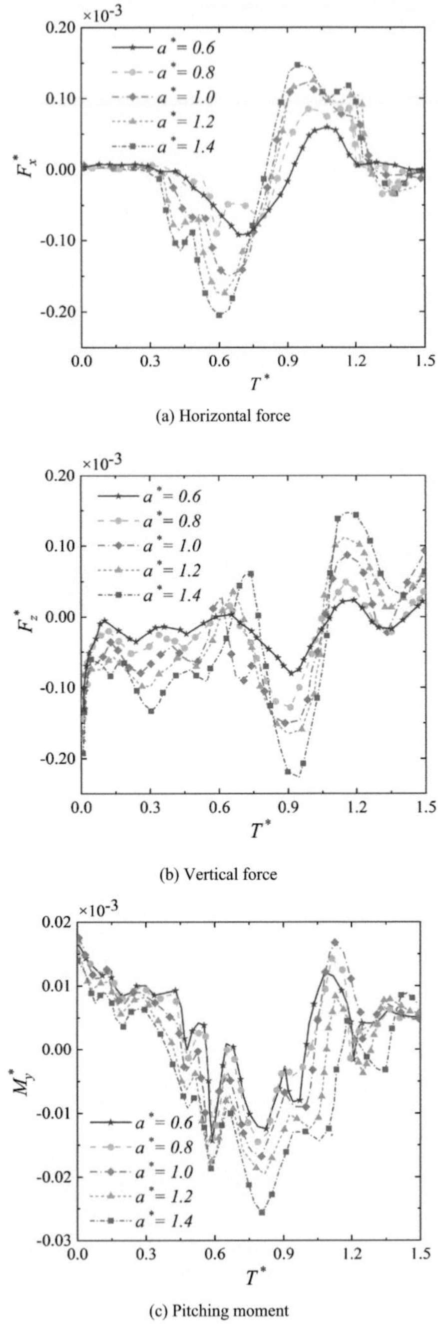

submerged body and the inertial force of the fluid, the submerged body initially experiences a downward oscillating vertical force. When the submerged body below the density interface reaches the left wave surface of the internal solitary wave, it is subjected to an upward vertical force from the fluid. The larger the amplitude of the internal solitary wave, the greater the peak upward vertical force. As the submerged body moves away from the internal solitary wave flow field, it begins to descend, and the vertical force gradually approaches zero.

Figure 5.7c shows the pitching moment variation curve of the suspended submerged body. It can be observed that the moment variation trends are generally consistent, with relatively intense moment oscillations. Initially, the submerged body experiences a bow-down moment, which reaches its peak when approaching the trough of the internal solitary wave. Larger wave amplitudes result in greater peak moments. As the submerged body moves away from the wave trough, the bow-down moment transitions to a bow-up moment, gradually approaching zero in magnitude.

5.3 Simulation of Interaction Between Internal Solitary Waves and Submerged Body with Speed

In addition to stationary suspension and hovering operations, underwater submerged bodies more frequently navigate freely at certain speeds. Generally, submerged bodies moving at speed are more prone to encountering hazardous situations when meeting internal solitary waves. Studying the interaction between freely moving submerged bodies and internal solitary waves is also crucial. Building upon the previous simulation of the interaction between suspended submerged bodies and internal solitary waves, this section further investigates the interaction between submerged bodies with speed and internal solitary waves. It simulates scenarios where submerged bodies encounter internal solitary waves during constant-speed free navigation and analyzes the influence patterns of internal solitary waves on the motion and forces of submerged bodies with speed.

5.3.1 *Influence of Submergence Depth*

To study the impact of submergence depth on the motion and forces of submerged bodies with speed under the action of continuously stratified internal solitary waves, the simulation maintains a constant internal solitary wave amplitude ($a = 50m$), a submerged body speed of 10 m/s, and a density interface thickness of 20 m. Five simulation cases (Case 11–15) with different submergence depths for submerged bodies with speed interacting with internal solitary waves were set up, with specific working conditions as shown in Table 5.3. When the internal solitary wave propagates to the working area of the computational domain ($D_\lambda = 1.1\lambda$), the three degrees of freedom of the submerged body with speed are released, and its surge, heave, and force load variations are monitored.

(1) Motion Response

Figure 5.8 presents the numerical calculation results of the motion response of submerged bodies with speed at different submergence depths. Since the horizontal speed of the submerged body is constant, the surge variation curve is a straight line, indicating uniform linear motion in the horizontal direction. The slope of the line represents the horizontal speed of the submerged body's navigation.

Figure 5.8b shows the heave motion curve of the submerged body. It can be observed that submerged bodies located below the pycnocline all sink downward after crossing the trough of the internal solitary wave, while those above or within the pycnocline initially sink and then begin to rise after encountering the left wave surface. Among them, the submerged body within the pycnocline rapidly rises upward after colliding with the left wave surface, as shown in Fig. 5.9. Figure 5.8c displays the center-of-mass trajectory curves of submerged bodies at different depths. It can be seen that the influence of continuous stratification on the submerged body traversing the internal solitary wave is smaller than that of strong stratification. The motion trajectory of the submerged body below the pycnocline is similar to that under strong stratification, but the drift rate toward the lower left is slightly slower.

(2) Load Characteristics

Figure 5.10 presents the numerical calculation results of the load characteristics of submerged bodies moving at speed under the influence of continuous stratification internal solitary waves at different depths. The horizontal force, vertical force, and pitching moment experienced by the submerged body are significantly smaller than those under strong stratification. Figure 5.10a shows the horizontal force variation curve of the submerged body. The primary horizontal force acting on the submerged body is the fluid resistance to the right. When the submerged body is released with degrees of freedom, its frontal area increases, leading to a sharp rise in horizontal force. After moving away from the influence range of the internal solitary wave field, the horizontal force begins to decrease.

Table 5.3 Working conditions table for submerged bodies with speed at different submergence depths under continuously stratified internal solitary waves

Case	Submergence depth d	Wave amplitude a	Speed V	Pycnocline thickness D_{pyc}
Case 11	+25 m ($d^* = 0.5$)	50 m	10m/s	20 m
Case 12	0 m ($d^* = 0$)			
Case 13	−25 m ($d^* = -0.5$)			
Case 14	−50 m ($d^* = -1.0$)			
Case 15	−75 m ($d^* = -1.5$)			

Fig. 5.8 Motion response of submerged bodies with speed at different submergence depths

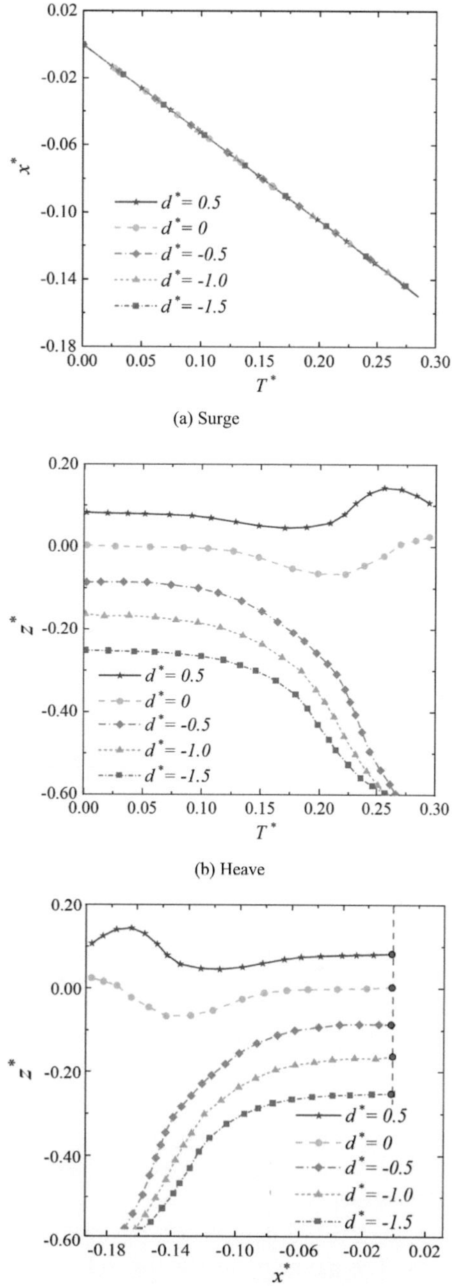

(a) Surge

(b) Heave

(c) Trajectory Curve of the Center of Gravity

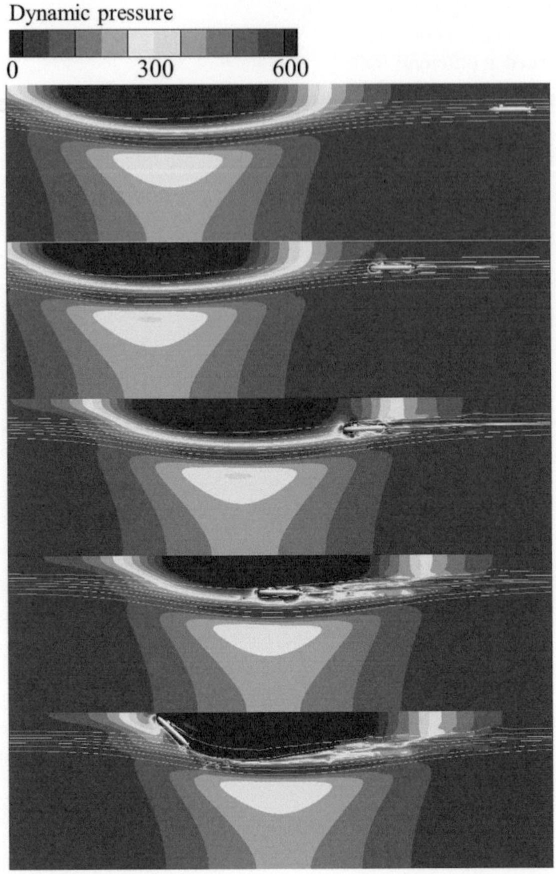

Fig. 5.9 Encounter between a submerged body moving at speed within the pycnocline and an internal solitary wave ($d^* = 0$)

Figure 5.10b illustrates the vertical force variation curve of the submerged body. The changes in vertical force for submerged bodies above and within the pycnocline are similar to those in horizontal force. As the submerged body approaches the right wave surface of the internal solitary wave, the downward vertical force increases. Upon collision with the left wave surface, the upward vertical force increases, causing the submerged body to rise. In contrast, for submerged bodies below the pycnocline, the vertical force continues to increase downward. After crossing the trough, the vertical force begins to increase upward, but due to the inertia of the submerged body, it still moves downward.

Figure 5.10c depicts the pitching moment variation curve of the submerged body. The moment changes for submerged bodies above the pycnocline are more pronounced. They first experience a bow-down moment, causing the submerged body to sink. After encountering the left wave surface, the moment direction reverses to bow-up, making the submerged body rise. Under the influence of the lower fluid on the stern, the submerged body begins to bow down again. The moment changes

Fig. 5.10 Load characteristics of submerged bodies moving at speed under different depths

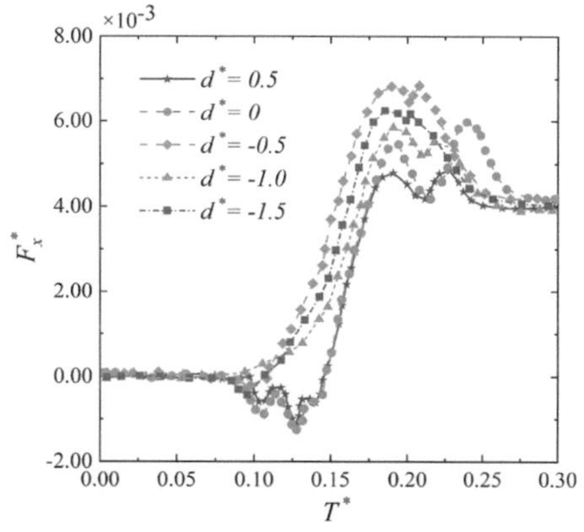

(a) Horizontal Force

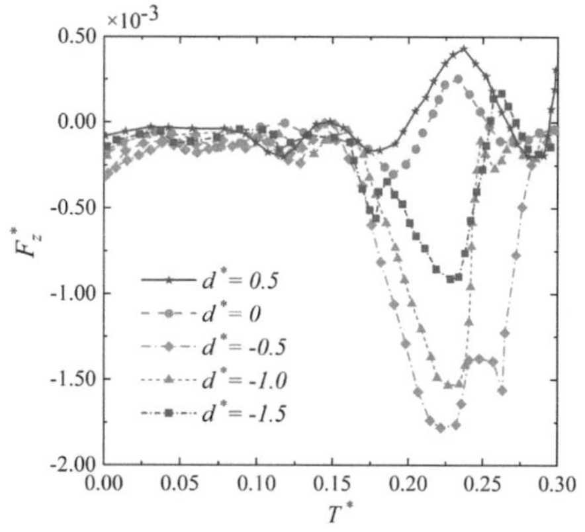

(b) Vertical Force

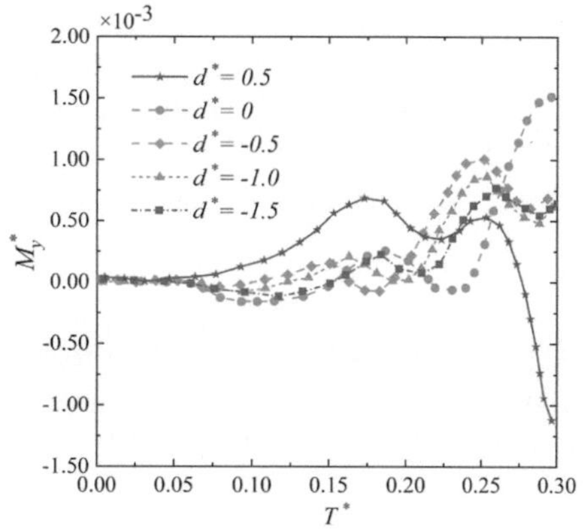

(c) Pitching Moment

for submerged bodies within and below the pycnocline are similar, but the moment reversal for submerged bodies within the pycnocline is slightly delayed because they traverse the pycnocline, while those below the pycnocline are directly affected by the flow field.

5.3.2 *Influence of Wave Amplitude*

To study the effect of internal solitary wave amplitude in continuous stratification on the motion and forces of a submerged body with speed, the simulation maintained a constant submergence depth (20 m below the interface), a body speed of 10 m/s, and a density interface thickness of 20 m. Five simulation cases (Case 16–20) were set up for the interaction between internal solitary waves of different amplitudes and the submerged body. The specific working conditions are listed in Table 5.4. When the internal solitary wave propagated into the working area of the computational domain ($D_\lambda = 1.1\lambda$), the three degrees of freedom of the submerged body were released, and its surge, heave, and force variations were monitored.

(1) Motion Response

Figure 5.11 shows the dynamic pressure field contours of the interaction between the submerged body with speed and internal solitary waves of different amplitudes. It can be observed that when the submerged body initially approached the right wave surface of the internal solitary wave, its motion attitude did not change significantly. However, the large-amplitude () internal solitary wave had a greater impact on the motion attitude of the submerged body, causing it to quickly pitch downward. During the propagation of the internal solitary wave, the left wave surface began to thin, and the flow field deformed. The larger the wave amplitude, the more pronounced the deformation trend. This is because the propagation speed of large-amplitude internal solitary waves is greater, and the left wave surface is squeezed under the combined effect of the lower fluid and the central fluid in the wave trough, leading to mixing at the density interface.

Table 5.4 Working conditions of a submerged body with speed under internal solitary waves of different amplitudes in continuous stratification

Case	Wave amplitude a	Submergence depth d	Speed V	Pycnocline thickness D_{pyc}
Case 6	30 m ($a^* = 0.6$)	−20 m	10m/ s	20 m
Case 7	40 m ($a^* = 0.8$)			
Case 8	50 m ($a^* = 1.0$)			
Case 9	60 m ($a^* = 1.2$)			
Case 10	70 m ($a^* = 1.4$)			

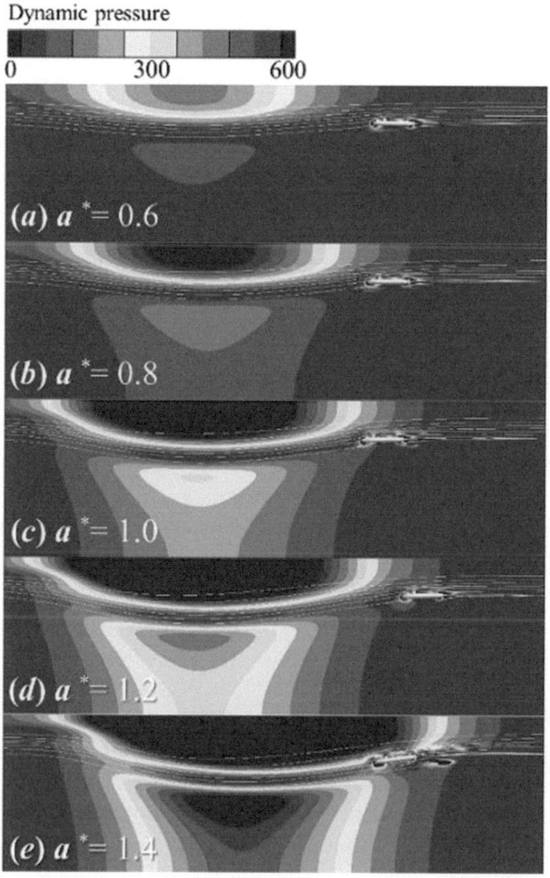

Fig. 5.11 Submerged body with speed located at the right wave surface of internal solitary waves with different amplitudes ($T^* = 0.1$)

Figure 5.12 shows the numerical calculation results of the motion response of a moving submerged body under internal solitary waves with different amplitudes in continuous stratification. Since the horizontal velocity of the submerged body is constant, its surge motion follows a straight line (as shown in Fig. 5.12a). The heave motion of the submerged body is illustrated in Fig. 5.12b. Similarly, except for the case of small wave amplitude ($a^* = 0.6$), where the moving submerged body does not cross the right wave surface of the internal solitary wave, it traverses the right wave surface under other amplitudes. When $a^* = 0.6$, the moving submerged body barely grazes the density interface, showing a tendency to pitch down but quickly recovers its normal attitude and continues sailing with minimal heave motion. Under the other four amplitudes, the moving submerged body pitches down and sinks, with the fastest descent speed observed at $a^* = 1.4$, indicating the greatest influence from the internal solitary wave. Figure 5.12c depicts the trajectory curves of the center of gravity of the moving submerged body under different amplitudes of internal solitary waves. It can be observed that wave amplitude significantly affects the motion response of the moving submerged body. Under small-amplitude internal solitary waves, the submerged body first pitches down and then up. As the wave amplitude increases, the moving submerged body continues to sink if no control is applied.

(2) Load characteristics

Figure 5.13 presents the numerical calculation results of the load characteristics of the moving submerged body under internal solitary waves with different amplitudes in continuous stratification. From Fig. 5.13a, it can be seen that as the moving submerged body approaches the right wave surface of the internal solitary wave, the horizontal force gradually increases, reaching its maximum at the trough of the internal solitary wave. Moreover, the peak horizontal force increases with larger wave amplitudes. Similar to the strong stratification case, under continuous stratification, the moving submerged body only traverses the internal solitary wave with small amplitude. Under large-amplitude internal solitary waves, the submerged body immediately pitches down and plunges into the "seabed," causing a sharp increase in horizontal force within a short time.

Figure 5.13b shows the variation curve of the vertical force on the moving submerged body. Due to the pressure difference on the upper and lower surfaces of the submerged body and the inertial force of the fluid, it initially experiences a downward vertical force. When reaching below the trough of the internal solitary wave, the downward vertical force reaches its minimum. After moving away from the trough, the submerged body is subjected to an upward force due to the lower fluid layer, reaching its maximum when located below the left wave surface. Beyond the trough, the vertical force oscillates more violently, and its magnitude is positively correlated with the wave amplitude.

Figure 5.13c shows the variation curve of pitching moment for the submarine with speed. The trend of pitching moment variation is similar under different wave amplitudes. When approaching the right wave surface of the internal solitary wave, the submarine experiences a bow-down moment, which transforms into a bow-up moment after crossing the wave trough. In the case of small wave amplitude ($a^* = 0.6$), the variation of pitching moment is relatively small.

5.3.3 *Influence of Navigation Speed*

Similar to the setting of steady motion without trim under strong stratification, with the thickness of density transition layer set to 10 m, we study the variation of load characteristics when a submarine with speed crosses continuous stratified internal solitary waves in the vertical plane. This section sets five different navigation speeds for the submarine to cross continuous stratified internal solitary waves in uniform linear motion (Case 21–25), with specific working conditions shown in Table 5.5.

Figure 5.14 shows the dynamic pressure cloud diagram of submarines with different speeds located at the trough of continuous stratified internal solitary waves.

Fig. 5.12 Motion response of the moving submerged body under different amplitudes

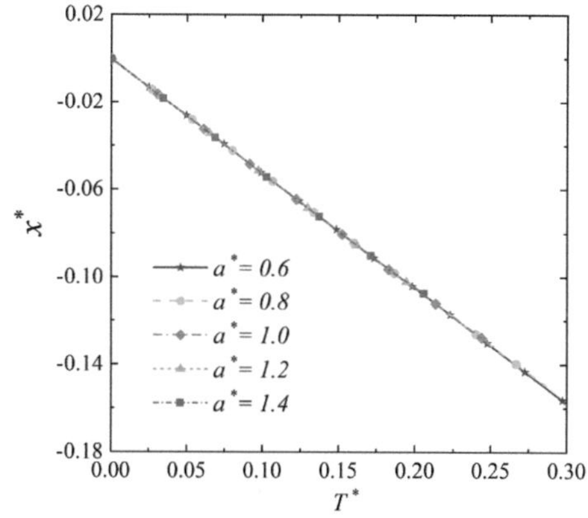

(a) Surge motion

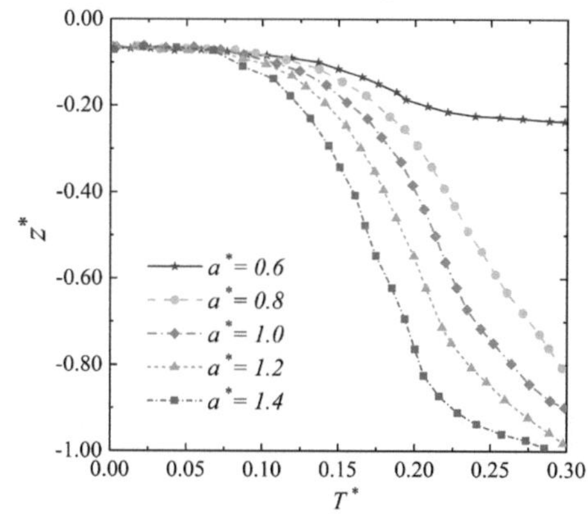

(b) Heave motion

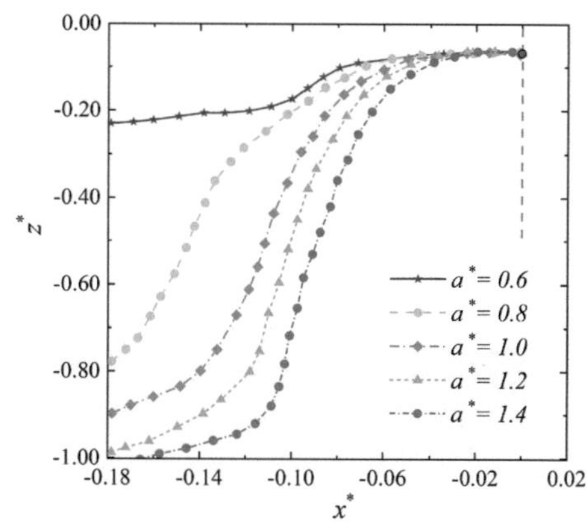

(c) Center of gravity trajectory

Fig. 5.13 Load characteristics of the submarine with speed under different wave amplitudes

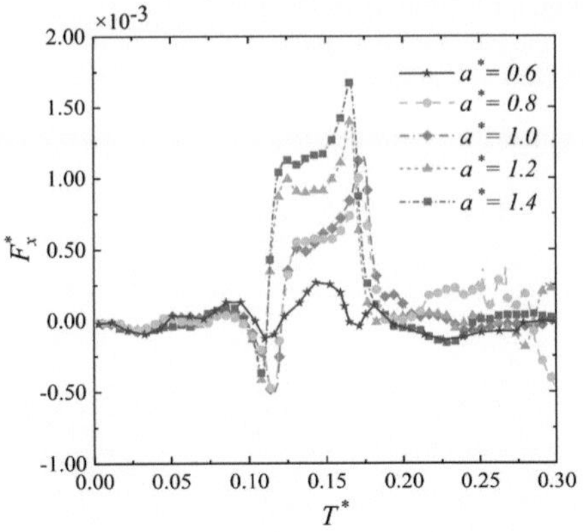

(a)Horizontal force

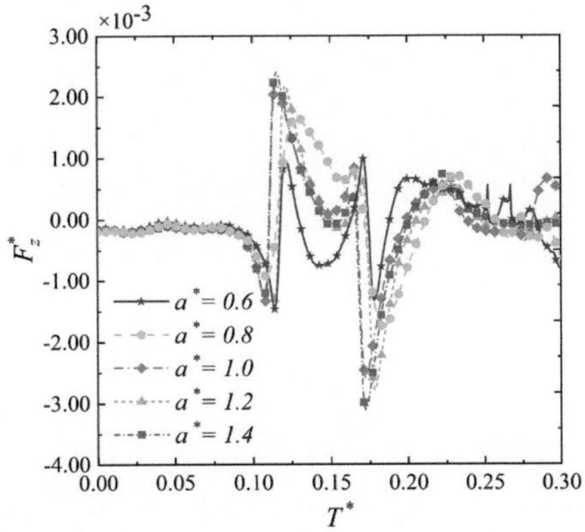

(b)Vertical force

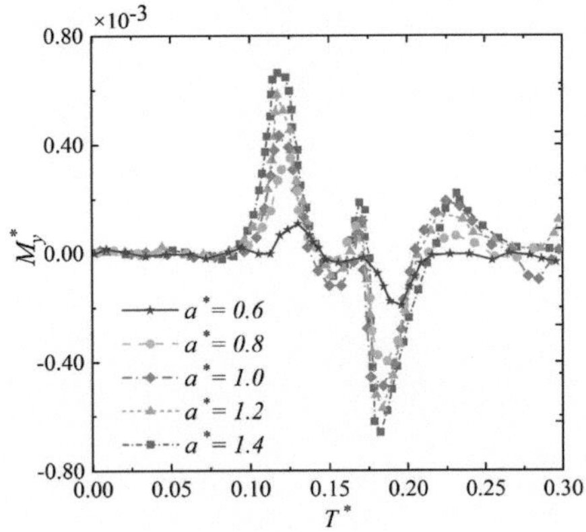

(c)Pitching moment

Table 5.5 Working condition table for submarines with different speeds

Case	Speed V	Submergence depth d	Wave amplitude a	Pycnocline thickness D_{pyc}
Case 21	5.0 m/s ($V^* = 0.171$)	−25 m	50 m	20 m
Case 22	7.5 m/s ($V^* = 0.256$)			
Case 23	10.0 m/s ($V^* = 0.342$)			
Case 24	12.5 m/s ($V^* = 0.427$)			
Case 25	15.0 m/s ($V^* = 0.513$)			

It can be observed that submarines with higher speeds leave more obvious wakes when crossing internal solitary waves, causing greater disturbance to the flow field.

Figure 5.15 shows the variation of load characteristics when submarines with different speeds cross continuous internal solitary waves. From the horizontal force variation curve in Fig. 5.15a, it can be seen that the trend of horizontal force variation is similar for submarines with different speeds when crossing internal solitary waves. When the horizontal degree of freedom is just released, the submarine moves in straight line at fixed speed. As it approaches the internal solitary wave, the horizontal force gradually increases; when the submarine contacts the right wave surface, the horizontal force reaches the first peak. During the process of crossing the wave trough, the horizontal force decreases, and reaches the second peak when arriving at the left wave surface. Different from the case of strong stratification, under strong stratification the horizontal force reaches its peak (only one peak) when the submarine arrives at the center of wave trough, and the peak value is larger than that under continuous stratification.

Figure 5.15b shows the vertical force variation curve. When the uniformly moving submarine passes through the internal solitary wave flow field, significant vertical force changes occur at both wave surfaces. Upon initially crossing the right wave surface from high-density to low-density water, buoyancy decreases while gravity remains constant, resulting in increased downward vertical force. When crossing the density interface layer, the vertical force fluctuates with greater impact at higher speeds. Upon reaching the left wave surface, under the influence of lower-layer fluid, the upward vertical force increases. While crossing the density interface layer, buoyancy increases and the vertical force begins decreasing toward zero.

Figure 5.15c displays the pitching moment curve, showing similar trends to the strongly stratified case. As speed increases, moment peaks grow while the internal solitary wave's duration of effect shortens. Initially, the submarine experiences a bow-down pitching moment. When passing the wave trough center, the moment direction reverses to bow-up and gradually increases. After leaving the wave, the

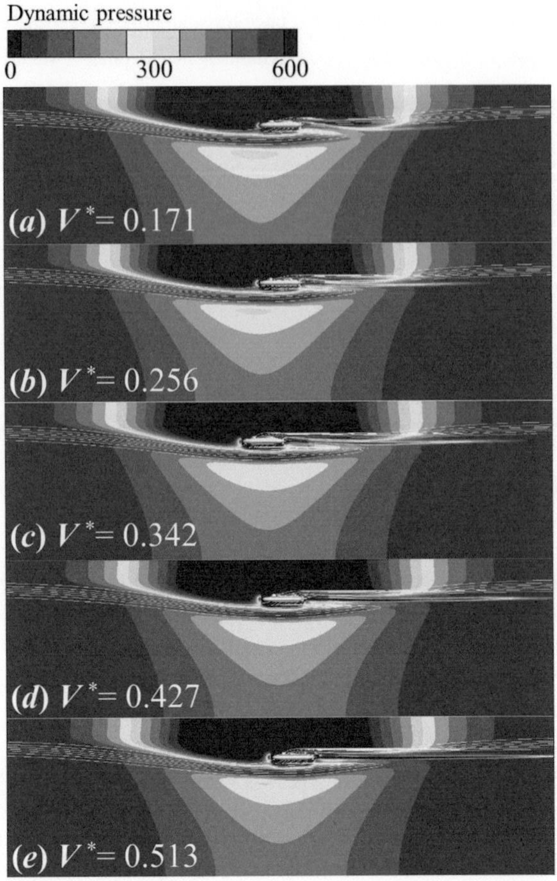

Fig. 5.14 Submarines with different speeds located at the trough of internal solitary waves

moment returns toward zero. The continuous stratification case produces greater moments than the strongly stratified case.

5.4 Summary

This chapter establishes a 3D continuous stratification internal solitary wave numerical flume based on the fully nonlinear theory—DJL equation combined with RANS $k-\varepsilon$ model using initialized flow field wave-making methods, investigating three key parameters (submergence depth, wave amplitude, and speed) on submarine motion responses and load characteristics. Research indicates:

Under the action of continuously stratified internal solitary waves, the submergence depth significantly affects the motion response of submerged bodies, especially for those near the pycnocline, which can cause substantial changes in the body's

Fig. 5.15 Load characteristics of submarines at different speeds

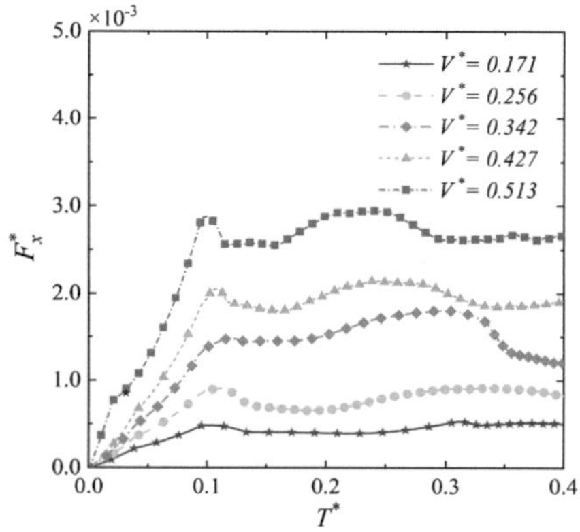

(a) Horizontal force

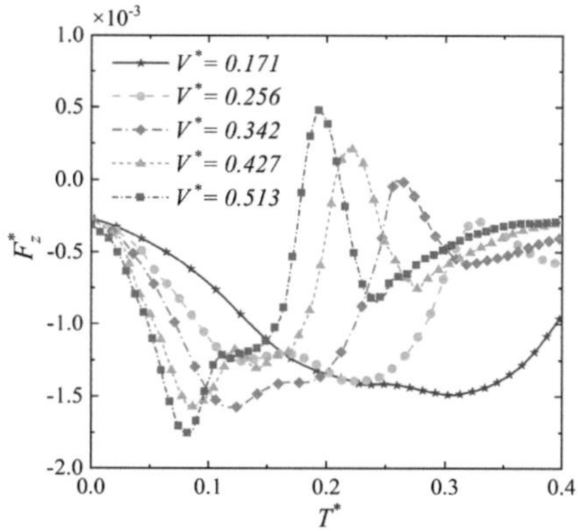

(b) Vertical force

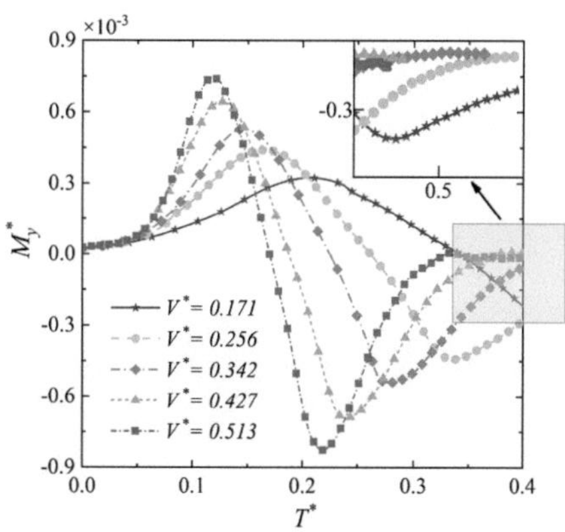

(c) Pitching moment

motion attitude, leading to abrupt load variations. Submerged bodies initially located within or above the pycnocline, whether in a suspended state or at constant speed, will experience vertical oscillations when interacting with continuously stratified internal solitary waves, resulting in drastic changes in vertical forces and pitching moments. In contrast, submerged bodies initially below the pycnocline will exhibit a downward nodding motion in both suspended and constant-speed states. The submergence depth () with the same wave amplitude corresponds to the maximum force amplitude. The motion responses of submerged bodies under internal solitary waves of different amplitudes are similar. Notably, a submerged body with forward speed can skim past the pycnocline of small-amplitude ($d^* = -0.5$) internal solitary waves without experiencing a "falling deep" phenomenon. The wave amplitude primarily influences the peak load magnitude and response speed—the larger the amplitude, the greater the peak load and the faster the response to the flow field. The forward speed mainly affects the load magnitude and the duration of wave-body interaction—the higher the speed, the greater the load and the shorter the interaction duration.

Furthermore, compared with conclusions drawn under strong stratification conditions, the presence of a pycnocline thickness causes oscillations in the motion and forces of submerged bodies, manifesting as slower response rates and abrupt changes in motion and forces. The thicker the pycnocline of the internal solitary wave, the more likely it is to cause waveform deformation during propagation, leading to mixing between upper and lower fluid layers.

Chapter 6
Influences of Stratified Structure and Internal Waves on Propeller

6.1 Introduction

Currently, surface ships and underwater vehicles primarily use propellers as propulsion devices. The hydrodynamic performance, cavitation, and noise characteristics of propellers have always been major concerns for ship researchers and designers. When a propeller operates behind a ship, its high-speed rotation creates a highly complex physical field, commonly referred to as the propeller wake field.

Depending on actual sea conditions, ship propellers often need to operate in complex and variable marine environments. Due to environmentally induced motions (e.g., waves, currents, internal waves, etc.), the operational characteristics of propellers can change significantly during actual operation. Propeller performance directly affects the propulsion efficiency and stealth capabilities of vehicles. Under the influence of ocean stratification and internal waves, propeller performance and its wake field may be adversely affected, potentially threatening the safe navigation of vehicles.

This chapter, based on the internal solitary wave numerical wave tank established earlier, studies the open-water performance of the INSEAN E1619 seven-blade right-handed propeller, which is commonly paired with SUBOFF. It analyzes its propulsion efficiency components and explores the effects of ocean stratification and internal waves on the hydrodynamic performance of the propeller.

P. Du et al., *Internal Wave Hydrodynamics on Submersibles*,
https://doi.org/10.1007/978-981-95-4479-0_6

Table 6.1 Main parameters of INSEAN E1619 propeller

Name	Parameter
Number of blades	7
Direction of rotation	Right-handed
Diameter (mm)	485
Hub-to-tip ratio	0.226
Pitch ratio $P0.7R$ (mm)	1.15
Chord length (0.75R) (mm)	6.8
Expanded area ratio A_E/A_O	0.608

6.2 Computational Model and Verification

6.2.1 Propeller Geometric Model

The propeller model used in this chapter's calculations is the INSEAN E1619 seven-blade right-handed propeller. This type of propeller is widely used in SUBOFF maneuverability numerical simulations. The INSEAN E1619 propeller's main parameters are listed in Table 6.1, and its 3D model schematic is shown in Fig. 6.1.

6.2.2 Computational Domain and Grid Independence Verification

To better simulate the interaction between internal waves and the propeller, the numerical simulation domain must avoid wall effects. The computational domain is divided according to ITTC recommendations: 12D upstream of the propeller, 50D downstream, and 24D in both lateral and depth directions (where D is the propeller diameter). During calculations, the propeller is set as a wall, with the inlet and outlet boundary conditions defined as velocity-inlet and pressure-outlet, respectively, while other boundaries are set as symmetry planes. The rotating region around the propeller is defined as a cylindrical moving zone, while the rest remains stationary, as shown in Fig. 6.2.

For meshing the propeller and numerical tank, a hybrid approach combining unstructured and structured grids is adopted. Due to the propeller's complex geometry, unstructured grids are used for the propeller itself, while structured grids are applied to stationary regions, as illustrated in Fig. 6.3.

From the perspective of numerical simulation accuracy, more grids generally yield more reliable results. However, excessive grids increase computational time. Under limited computational resources, a balance must be struck—insufficient grids lead to inaccuracies. Therefore, grid refinement is applied to regions with complex flows and high velocity gradients (e.g., blade tips, roots, leading/trailing edges, and

Fig. 6.1 3D model of the INSEAN E1619 propeller

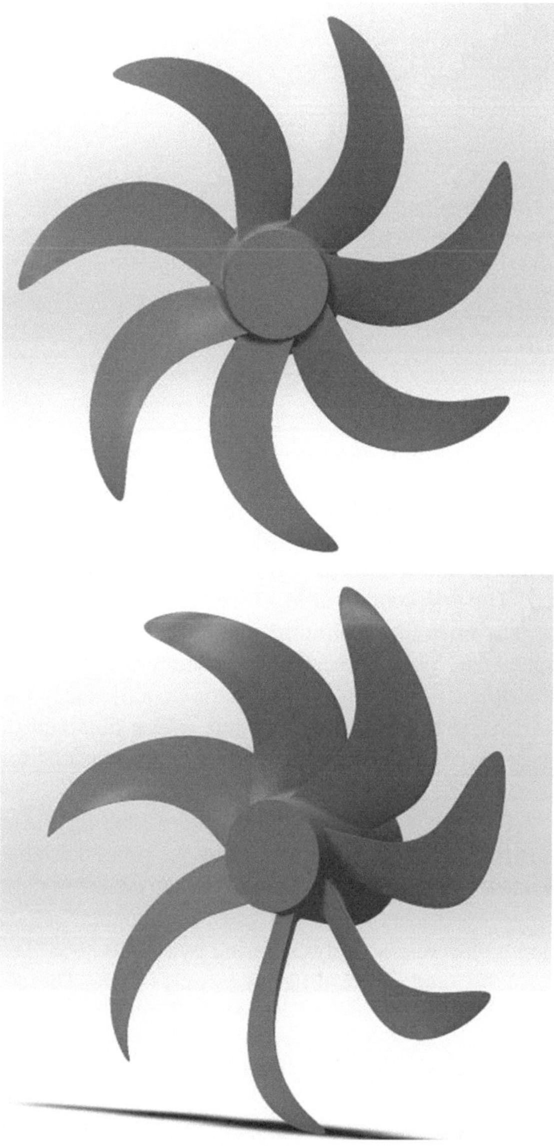

surrounding flow fields), as shown in Fig. 6.4, while coarser grids are used in less dynamic regions to control the total grid count.

Based on this, grid independence verification was conducted, as the error caused by format viscosity gradually increases with grid refinement, and accuracy can no longer be improved. During the grid independence verification, three sets of grids were generated: coarse grid, medium grid, and fine grid. The number of background grids increases with the number of foreground grids, and the grids in overlapping

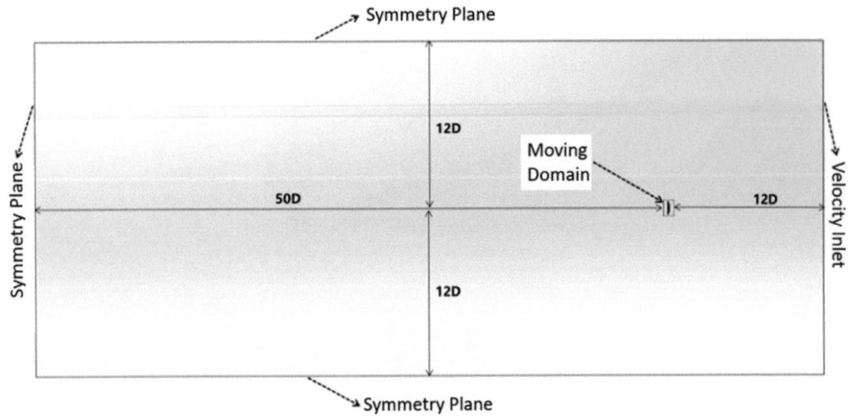

Fig. 6.2 Schematic of the open-water experiment computational domain

regions are refined to ensure a smooth transition between foreground and background grids. In non-overlapping regions, the grids gradually widen with a growth rate of 1.2. The grid counts for the three sets are listed in Table 6.2.

The operating condition was set as inflow velocity VA = 1.455 m/s, propeller speed n = 5 r/s. Two calculated parameters were selected for comparative analysis: the thrust coefficient K_T and torque coefficient 10 K_Q generated by the propeller, which are the most fundamental values obtained during the calculation of open water performance curves. The calculation results for the three grid sets are shown in Table 6.3.

As can be seen from Table 6.3, under medium grid conditions, the thrust coefficient and torque coefficient produced by the propeller have basically stabilized, and their values no longer change drastically with increasing grid numbers. The errors between calculated and experimental thrust coefficients for medium and fine grids are 10.76% and 9.99% respectively, differing by only 0.86%. The errors for torque coefficients are 1.38% and 1.63%, differing by only 0.25%. Therefore, it can be concluded that the medium grid can already accurately predict the propeller's performance parameters.

6.2.3 *Open Water Performance Verification Analysis*

To verify that the numerical simulation method adopted in this paper can accurately simulate the propeller's performance parameters, this section conducts numerical simulation of open water tests for the INSEAN E1619 seven-blade right-handed propeller and compares the results with measurements from the Italian CNR-INSEAN water tank experiments. During the calculation process, the inflow velocity VA, propeller speed n, and advance coefficient J are defined and used as initial conditions for the fluid domain, where the definition of advance coefficient J is given by

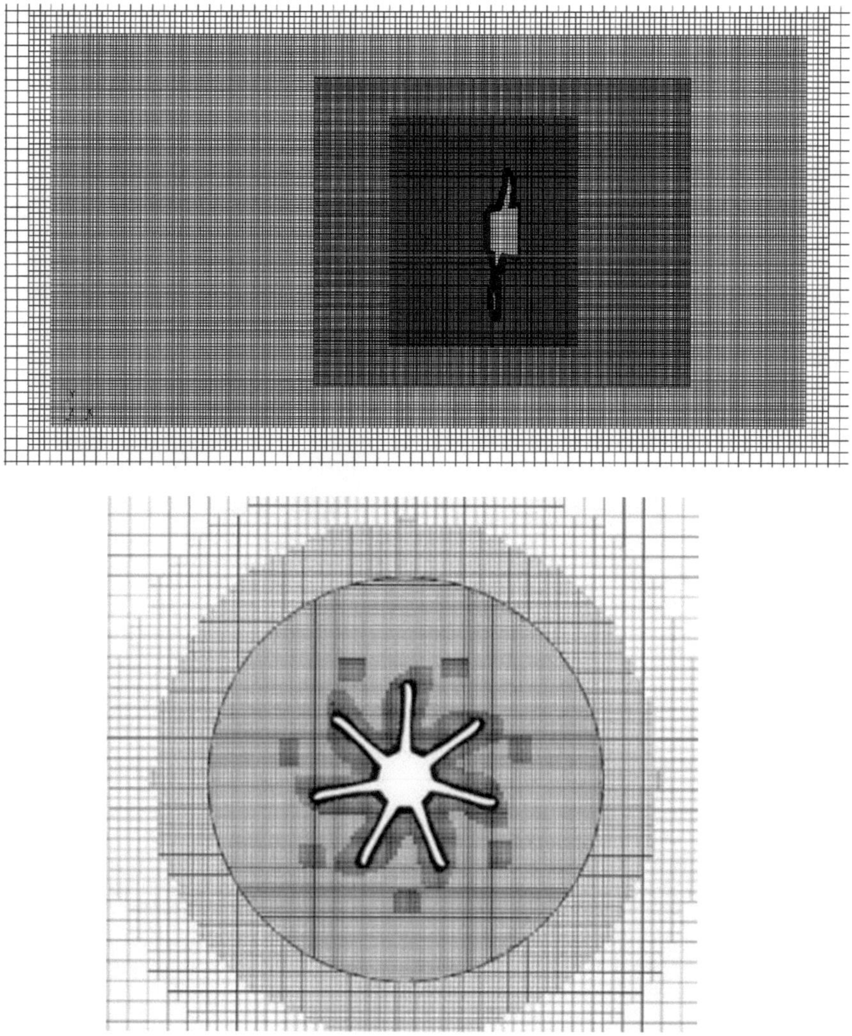

Fig. 6.3 Computational domain and propeller mesh partitioning

Eq. (6.1):

$$J = \frac{V_A}{nD} \tag{6.1}$$

where D is the propeller diameter.

During the calculation process, the initial value of advance coefficient J is 0. The J value increases by 0.2 every 501 iterations. For each J value, this number of iterations

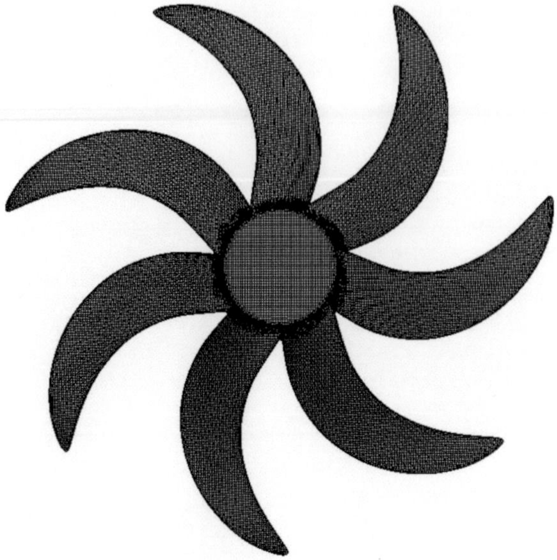

Fig. 6.4 Propeller surface mesh partitioning

Table 6.2 Grid independence verification

Grid type	Coarse grid	Medium grid	Fine grid
Foreground grid	457,039	994,554	1,806,728
Background grid	608,039	2,083,148	3,064,576
Total grid count	1,065,078	3,077,702	4,871,304

Table 6.3 Thrust coefficient and torque calculated from three grid sets

Grid type	Coarse grid	Medium grid	Fine grid
Calculated thrust coefficient KT	0.2465	0.2903	0.2928
Error relative to experimental value	24.2%	10.76%	9.99%
Calculated torque coefficient 10KQ	0.4461	0.5211	0.5199
Error relative to experimental value	15.5%	1.38%	1.63%

is sufficient to achieve solution convergence. When the J value changes, the inflow velocity V_A changes accordingly.

For a propeller with fixed geometry, its thrust coefficient, torque coefficient and efficiency are only related to the advance coefficient J. Their calculation formulas are shown in Eqs. (6.2), (6.3), and (6.4) respectively:

$$K_T = \frac{T}{\rho n^2 D^4} \tag{6.2}$$

$$K_Q = \frac{Q}{\rho n^2 D^5} \tag{6.3}$$

$$\eta_0 = \frac{K_T}{K_Q}\frac{J}{2\pi} \tag{6.4}$$

where T is the thrust of the propeller, and Q is the torque of the propeller.

The open water performance curve of the propeller refers to the relationship between K_T, K_Q, η_0 and J, where the abscissa is the advance coefficient J, and the ordinates are K_T, K_T, and η_0 respectively. The simulation adopts the SST k-ω turbulence model, with second-order spatial discretization and first-order temporal discretization for solving, Using the SIMPLE (Semi-Implicit Method for Pressure-Lined Equation) algorithm to solve the velocity–pressure coupling, and the SST (Shear Stress Transport) k-ω turbulence model is used to calculate the hydrodynamic and flow field information of the propeller.

Figure 3.5 shows the propeller open water curves calculated by the above simulation method compared with the experimental open water curves, where the smooth curves represent the values measured by the towing tank experiment, and the scattered points represent the values calculated by numerical simulation.

From Fig. 6.5, it can be seen that the results obtained by CFD calculation are basically consistent with the experimental results. In the range of $0.2 \leq J \leq 1$, the errors of CFD simulation results are relatively small, with the maximum error of thrust coefficient within 5% and the maximum error of torque coefficient within 10%. When the advance coefficient $J \leq 0.6$, the thrust coefficient and torque coefficient obtained by CFD calculation are both smaller than the experimental results, while when $J \geq 0.8$, the thrust coefficient and torque coefficient obtained by CFD calculation gradually exceed the experimental results. The open water efficiency of the propeller is closely related to thrust and torque, so its trend is also roughly consistent with the above conclusions. Compared with the experimental values, the errors between the CFD calculation values and the experiments are all within acceptable ranges, so it can be considered that the CFD method adopted in this chapter can accurately predict the open water performance of the propeller, which lays a good foundation for subsequent research on the influence of stratified ocean structures and internal waves on the hydrodynamic performance of the propeller.

6.3 Research on Hydrodynamic Characteristics of Propeller Under Stratified Structure

For the actual ocean, the density is often continuously changing, which is usually simplified into three parts: the upper uniform density layer, the middle strong density gradient/transition layer, and the lower uniform density layer.

Based on the previous section, this section establishes a density-stratified flow simulation method based on the thermal buoyancy-driven flow model. By specifying the temperature distribution in the flow field and the functional relationship between density and temperature, continuous density stratification is achieved. Assuming

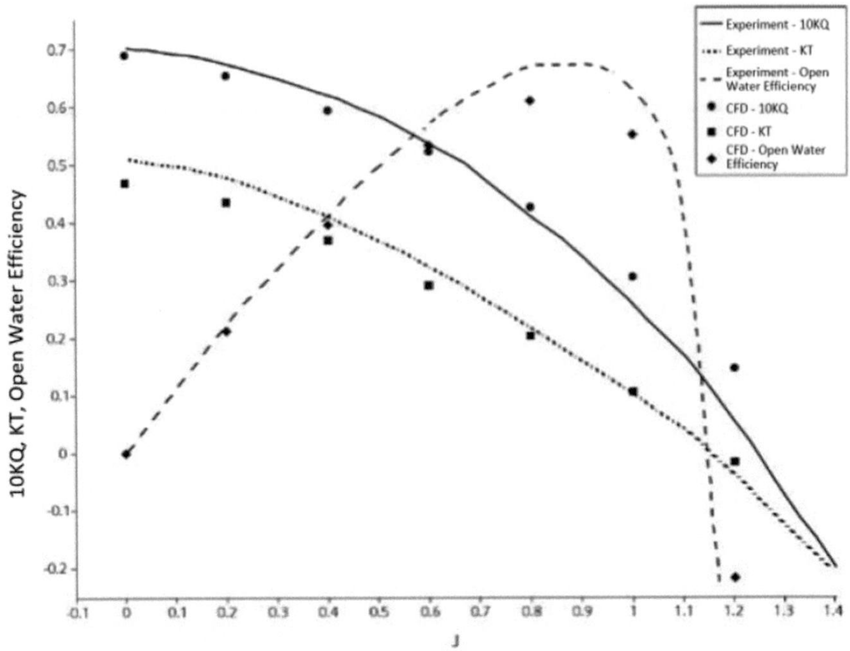

Fig. 6.5 Open water performance curves of the propeller obtained by CFD and experiments

that temperature changes caused by heat conduction during numerical processes are minimal, a numerical simulation method for continuously stratified density flow is established to investigate the influence laws of density stratification structure on propeller performance.

6.3.1 *Continuous Ocean Stratification Model*

For seawater in actual oceans, in open sea areas, salinity can be considered as stably distributed, and the influence of pressure on seawater density can be neglected. Therefore, temperature becomes one of the important factors affecting seawater density distribution. The buoyancy-driven flow based on temperature is called thermal buoyancy-driven flow. This paper assumes that fluid density is only related to temperature, thus establishing continuously stratified density fluid based on the thermal buoyancy-driven flow model. This is achieved by specifying the temperature distribution in the flow field and the functional relationship between density and temperature to realize continuous vertical density stratification.

Compared with simulations in homogeneous density fluids, simulating density-stratified flow requires additional solution of the energy equation (Eq. 2.3), written in integral form (Eq. 3.5). Typically, heat transfer processes include conduction, convection, and radiation. In this simulation, the influence of thermal radiation is neglected, considering only the effects of heat conduction and convection on the flow field. It is also assumed that the heat conduction coefficient is sufficiently small over the simulation time span to maintain stable density stratification, meaning heat conduction can also be neglected.

$$\frac{\partial}{\partial t}\int_{V}\rho E dV+\oint_{A}\rho H\boldsymbol{v}\cdot d\boldsymbol{a}=-\oint_{A}\boldsymbol{q}\cdot d\boldsymbol{a}+\oint_{A}\boldsymbol{T}\cdot\boldsymbol{v}d\boldsymbol{a}+\oint_{V}\boldsymbol{f}_{b}dV \tag{6.5}$$

Since only heat conduction and convection are considered, the flux q in the above equation only includes the heat conduction term (6.6) and heat convection term (3.7).

$$\boldsymbol{q}=-k\nabla T \tag{6.6}$$

$$\boldsymbol{q}=h(T_s-T_{ref}) \tag{6.7}$$

At this point, Eq. (6.5) can be transformed into a function of temperature. By taking temperature as the solution variable and writing fluid density as a function of temperature, continuous linear density stratification can then be achieved by specifying either linear vertical temperature distribution or piecewise functions.

The computational domain used in this chapter remains consistent with the settings in Sect. 2.2 regarding computational domain and boundary conditions. To achieve ocean stratification structure, the vertical temperature distribution expression in the flow field is first specified as shown in (6.8), and the functional relationship between density and temperature is defined as shown in (6.9):

$$T=T(y)=T_0+y \tag{6.8}$$

$$\rho=\rho(T)=g(y)=\rho_0-c*y \tag{6.9}$$

where $\rho(T)$ is a linear function, T_0 is the average temperature of the stratified ocean (in this section T_0= 300.12 K), ρ_0 is the average density of the stratified ocean (in this section $\rho_0 = 1024$ kg/m^3), and c is a constant whose value is set according to specific working conditions (Figs. 6.6, 6.7 and 6.8).

Setting the average temperature as $T_0 = 300.12\,K$ and the average density of the stratified ocean as $\rho_0 = 1024$ kg/m^3, the initialized flow field temperature and density distributions are shown in Figs. 3.7 and 3.8.

6.3.2 *Analysis of Stratification Effects*

Set the density variation range of the ocean stratification structure from 1020 kg/m^3 to 1028 kg/m^3, with all other settings consistent with the simulation method of

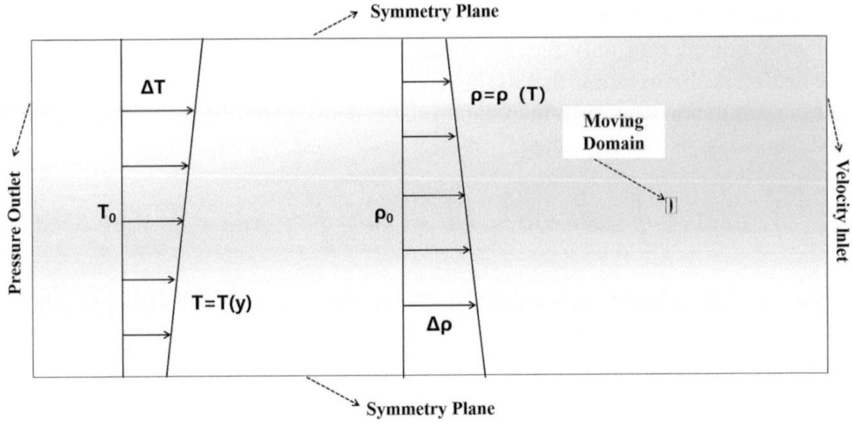

Fig. 6.6 Geometric model of ocean stratification structure

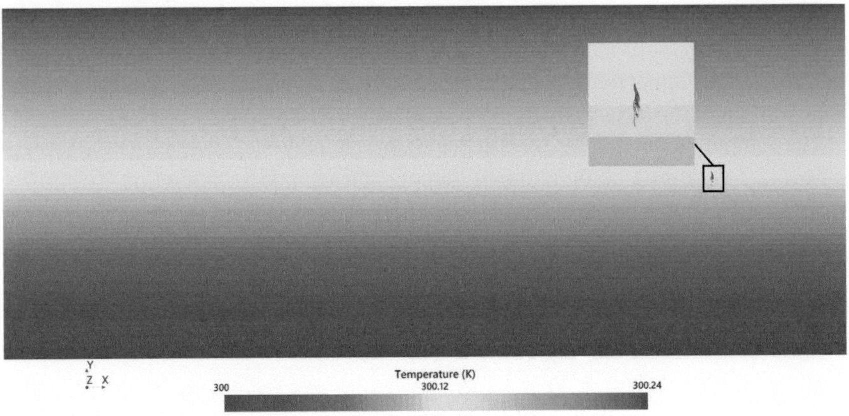

Fig. 6.7 Vertical temperature distribution contour of flow field

the open-water experiment. During the calculation, the inflow velocity V_A, propeller rotation speed n, and advance coefficient J are defined and used as initial conditions for the fluid domain.

During the calculation, the initial value of J is 0. J increases by 0.2 every 501 iterations. For each J value, this iteration count is sufficient to ensure convergence. When the J value changes, the inflow velocity V_A also changes accordingly.

Based on the calculation results, the open-water performance curves of the propeller in the stratified ocean structure are plotted and compared with the open-water curves of the propeller in a uniform fluid obtained in Sect. 2.3. The horizontal axis represents the advance coefficient J, while the vertical axes are K_T, $10K_Q$, and η_0, respectively. Figure 6.9 shows the propeller performance curves calculated in the stratified ocean structure compared with the open-water curves obtained in a uniform

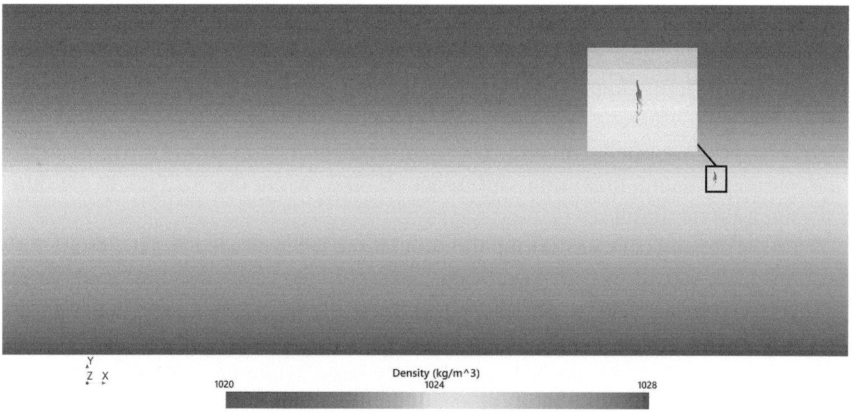

Fig. 6.8 Vertical density distribution contour of flow field

flow, including three curves. The smooth curve represents the values obtained in a non-stratified structure, while the scattered points represent the values calculated in the stratified ocean structure.

From Fig. 6.9, it can be observed that the propeller thrust coefficient, torque coefficient, and open-water efficiency calculated in the stratified ocean structure

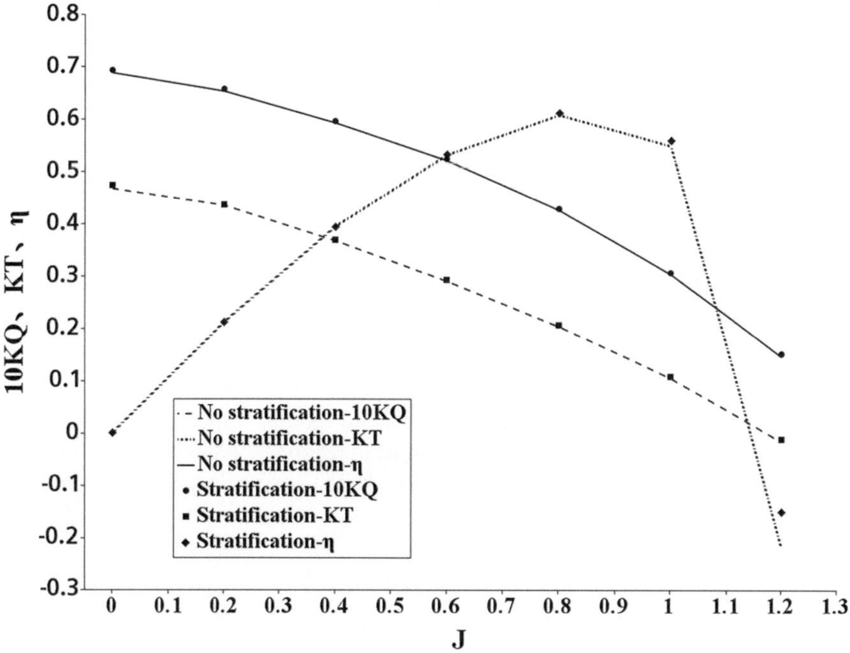

Fig. 6.9 Open-water curves of the propeller in stratified and non-stratified ocean structures

exhibit an extremely high degree of agreement with the simulation results from the non-stratified structure. When the advance coefficient $J \leq 1.0$, the values almost completely overlap; when $J \geq 1.0$, slight differences emerge, with the thrust coefficient, torque coefficient, and open-water efficiency being marginally higher than those in the non-stratified structure. The open-water efficiency shows the most significant variation, with a maximum difference of 30%, while the changes in the other two values are negligible.

From the density contours during the simulation, it can be seen that the propeller's motion significantly disturbs the surrounding stratified ocean structure, as shown in Fig. 6.10. The disturbance is greatest when the propeller begins rotating and gradually stabilizes over time. Under steady disturbance conditions, the propeller's lateral disturbance range in the wake field is relatively large, while the vertical disturbance range is smaller.

To avoid the scenario where the small density variation range of the stratified structure makes it difficult to observe its impact on propeller performance, the density variation range was increased. The density variation range of the stratified ocean structure was set from 1000 kg/m^3 to 1050 kg/m^3, while keeping other parameters unchanged, the motion of the propeller in the stratified structure was simulated again, with the density distribution cloud diagram shown in Fig. 6.11.

Figure 6.12 shows the propeller performance curves calculated in the ocean stratified structure with expanded density range compared with the open-water curves obtained in the non-stratified structure. It can be seen that the conclusions are consistent with those obtained in the small density range.

Based on the above simulation results, it can be concluded that the ocean stratified structure has little effect on the hydrodynamic performance of the propeller. Only when the advance coefficient $J \geq 1.0$, the thrust coefficient, torque coefficient, and open-water efficiency are slightly higher than those in the non-stratified structure, with the effect on open-water efficiency being the most significant.

6.4 Study on Hydrodynamic Characteristics of Propeller Under Internal Wave Action

Internal solitary waves are an important factor affecting the design and normal operation of navigational bodies. The load effects and motion responses caused by the interaction between internal solitary waves and navigational bodies cannot be ignored for the stability and operability of structures. As the power device of navigational bodies, the effect of internal waves on propellers will greatly affect the navigation efficiency and safety of navigational bodies. Therefore, exploring the interaction mechanism between internal waves and propellers is particularly important.

Based on Sect. 6.3, this section generates continuous stratified internal solitary waves in the numerical wave tank based on DJL theory and investigates the hydrodynamic characteristics of propellers under internal solitary wave action.

Fig. 6.10 Density contours during propeller motion

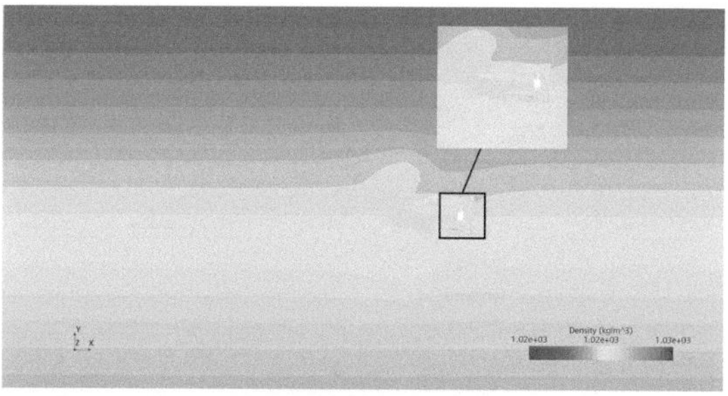

(a) density contour at 0.2s

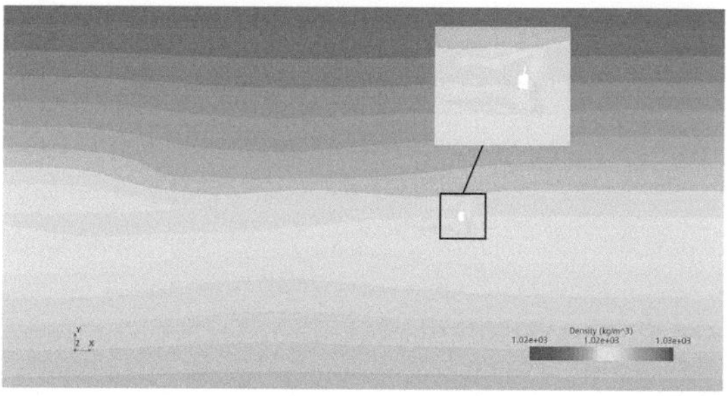

(b) Density contour at 1s

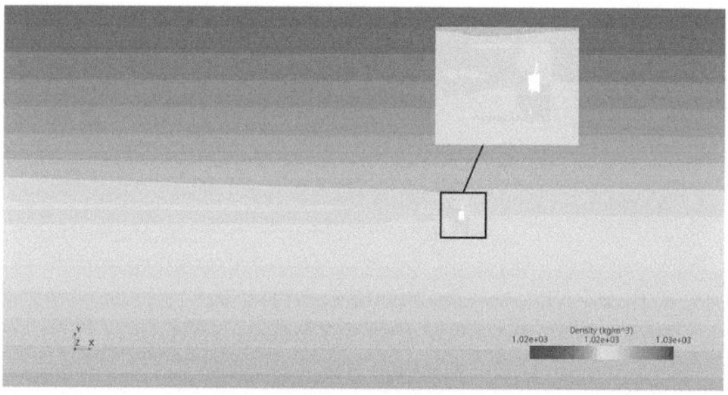

(c) density contour at 2s

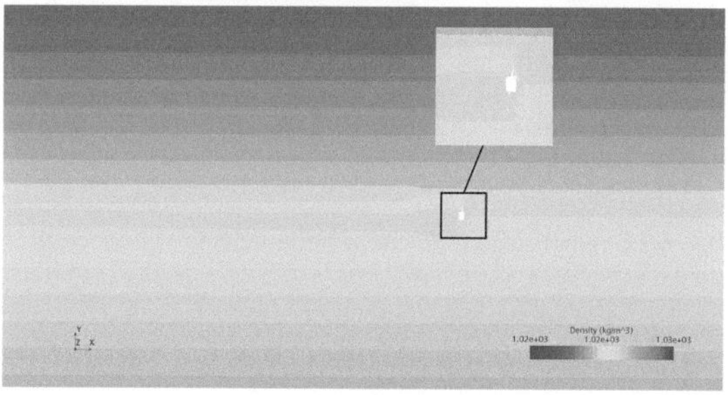

(d) Density contour at 4s

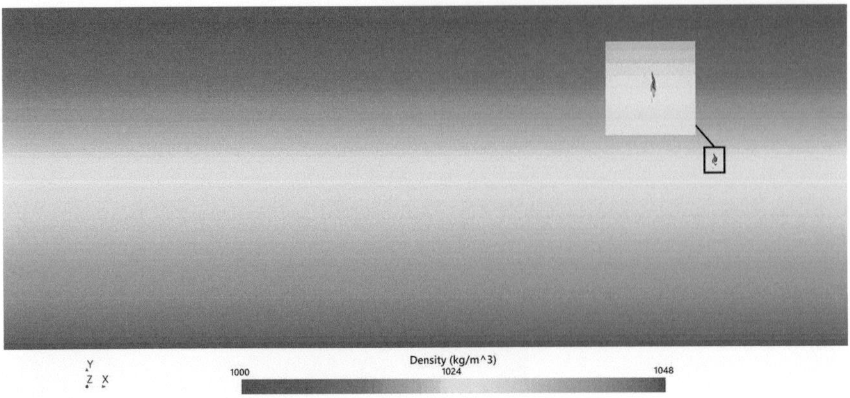

Fig. 6.11 Initial flow field after expanding the density range

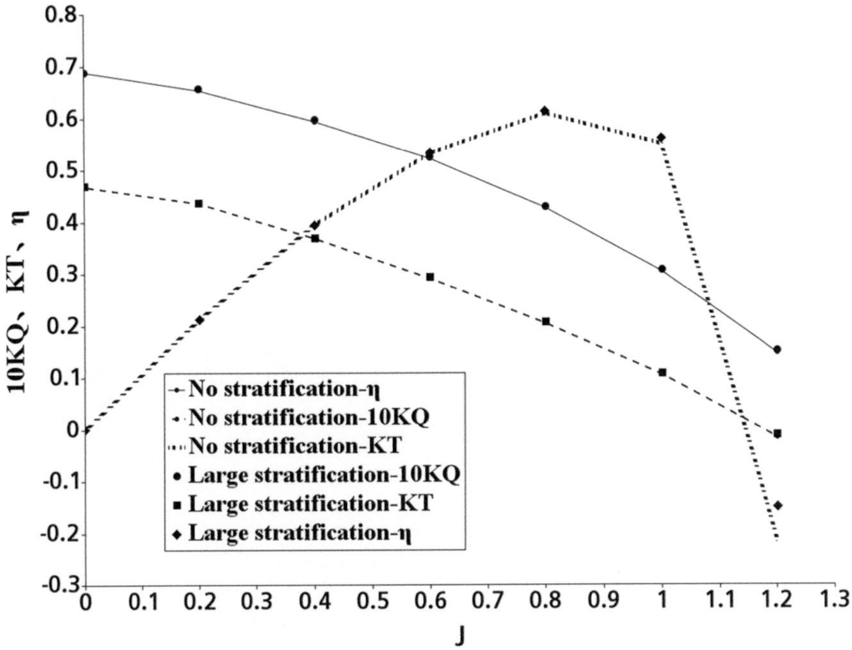

Fig. 6.12 Open-water curves in large-density-range stratified structure and uniform flow

6.4.1 Geometric Model and Computational Domain

In real oceans, the amplitude of internal solitary waves is generally much larger than that of surface waves, reaching up to hundreds of meters. The wavelength of internal

Table 6.4 Parameters of E1619 propeller scaled by 0.1

Name	Parameter
Number of blades	7
Direction of rotation	Right-handed
Diameter (mm)	48.5
Hub-to-tip ratio	0.226
Chord length (0.75R) (mm)	0.68
Expanded area ratio A_E/A_O	0.608

solitary waves is generally several hundred meters, and can even reach over ten thousand meters at maximum. Therefore, to ensure that the scale of internal waves relative to the propeller conforms to real-world conditions, the propeller model was scaled down by a factor of 0.1, with specific parameters shown in Table 6.4. Meanwhile, to provide stable propagation space for internal waves and ensure the integrity and stability of internal wave waveforms when encountering the propeller, the computational domain from the previous two sections was expanded while keeping other conditions unchanged. The numerical tank computational domain model used in this section is shown in Fig. 6.13.

The dimensions of the internal solitary wave numerical tank are 80 m in length and 10 m in water depth. The upper fluid layer has a thickness $h_1 = 2.0$ m and density $\rho_1 = 1020$ kg/m^3, while the lower fluid layer has a thickness $h_2 = 8.0$ m and density. The density transition layer thickness $h_0 = 0.1$ m.

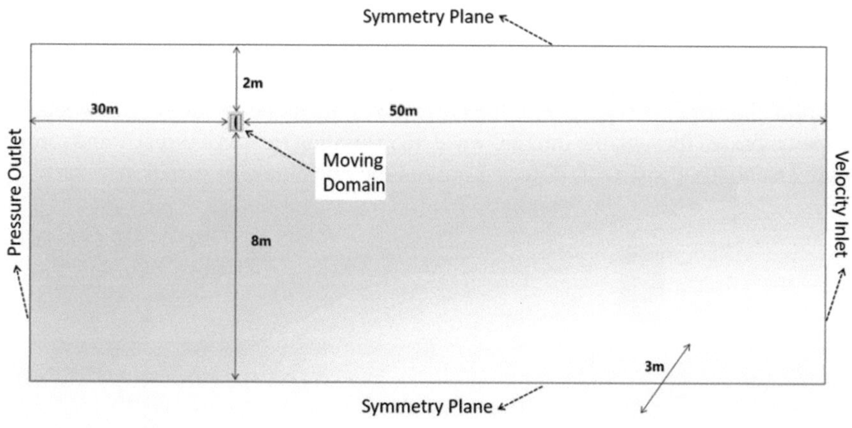

Fig. 6.13 Schematic diagram of computational domain for interaction between internal waves and propeller

6.4.2 Hydrodynamic Characteristics of a Propeller Crossing an Internal Solitary Wave

In this simulation, the interpolation wave-making method is used to generate internal solitary waves. Based on the DJL model, a program is written to select appropriate internal wave parameters and solve for the corresponding theoretical solution of the internal wave. The obtained parameters of the internal wave, such as density, velocity, pressure, and volume fraction, are directly interpolated into the numerical flume to establish the initial flow field, thereby realizing the generation of internal waves. Numerical dissipation is applied at the tail of the flume to eliminate wave reflections and avoid interference from trailing waves. During mesh generation, the flow field around the propeller and near the internal wave surface is refined to better capture the details of the interaction between the internal wave and the propeller.

In the standard test case, the internal wave has a height of 1.48 m and a wavelength of 30 m. The internal wave surface is located 1 m below the upper interface, while the propeller is positioned 2 m below the upper interface, allowing the propeller to precisely pass through the internal wave surface. The initial position of the internal wave (based on the trough position) is located 20 m from the inlet and 35 m from the propeller, ensuring sufficient space for stable wave propagation, as illustrated in Fig. 6.14.

Since the propagation of internal solitary waves induces a clockwise-rotating velocity field in the flow, no additional initial flow velocity is required. Only the rotational speed of the propeller needs to be applied, with the propeller speed n set to 10r/s. During the simulation, the thrust T and torque Q generated by the propeller are monitored to investigate the influence of the internal wave on the hydrodynamic characteristics of the propeller as it passes through. The obtained results are shown in Fig. 6.15.

From an overall perspective of the two figures, it can be observed that when the propeller passes through the internal wave, the variation trends of thrust T and torque Q are essentially consistent, differing only in the magnitude of variation.

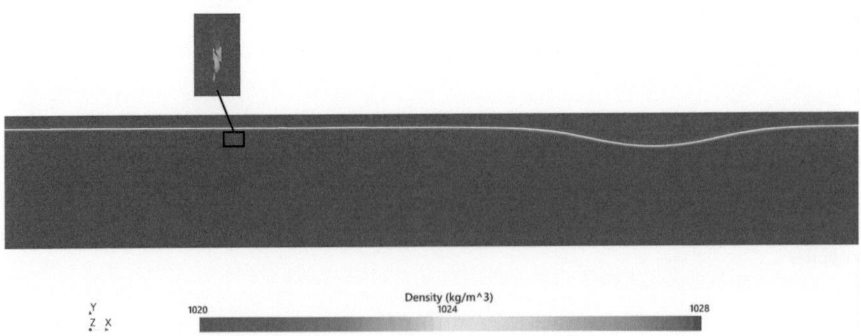

Fig. 6.14 Initial flow field contour

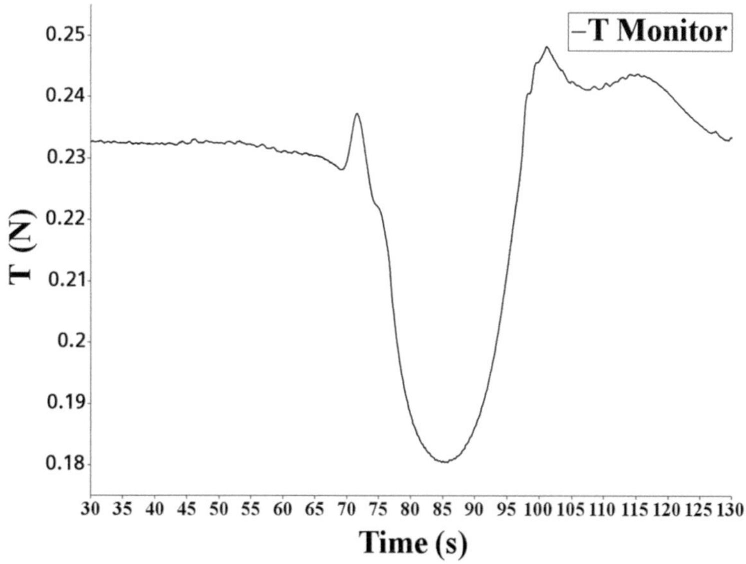

(a) Thrust T variation curve

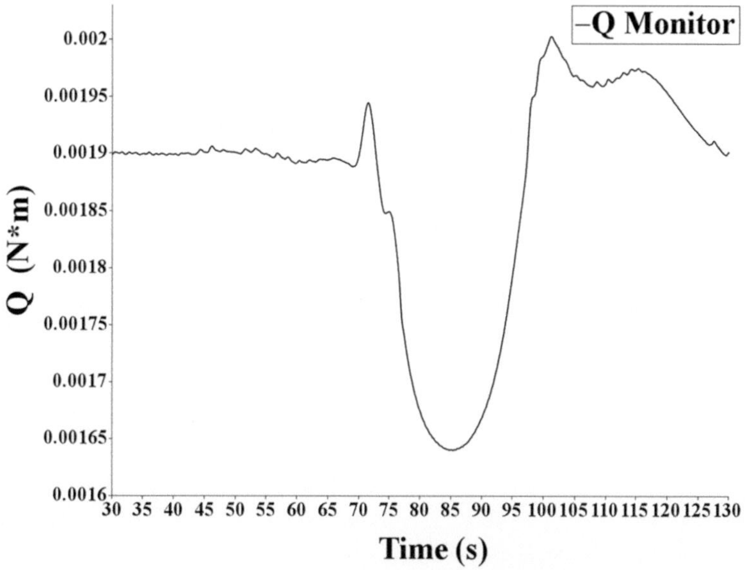

(b) Torque Q variation curve

Fig. 6.15 Variation curves of propeller performance parameters under internal solitary wave action

From the variation of thrust T, when the internal wave is far from the propeller, the thrust generated by the propeller remains relatively stable, maintaining around 0.232N. As the internal wave gradually approaches but has not yet contacted the propeller (i.e., between 55 and 69 s in Fig. 6.19), the thrust on the propeller slightly decreases. At this point, the trough center of the internal wave moves to a position 40 m from the inlet and 10 m from the propeller. The density contour around the propeller at this stage is shown in Fig. 6.16. By 69 s, thrust T drops to its lowest point, approximately 0.228N, which coincides with the moment the propeller begins to contact the internal wave surface, as illustrated in Fig. 6.17.

Starting from 69 s, thrust T begins to rise gradually, reaching a local maximum around 72 s. The density contours at this stage are shown in Figs. 3.18 and 3.19. These

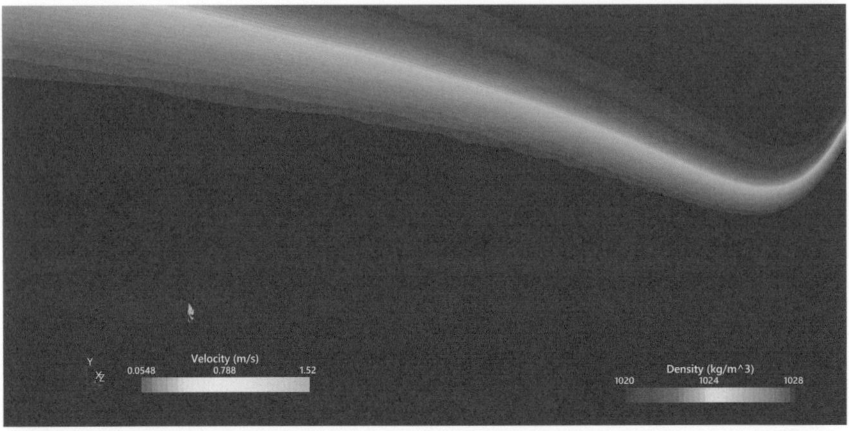

Fig. 6.16 Density contour at 60 s

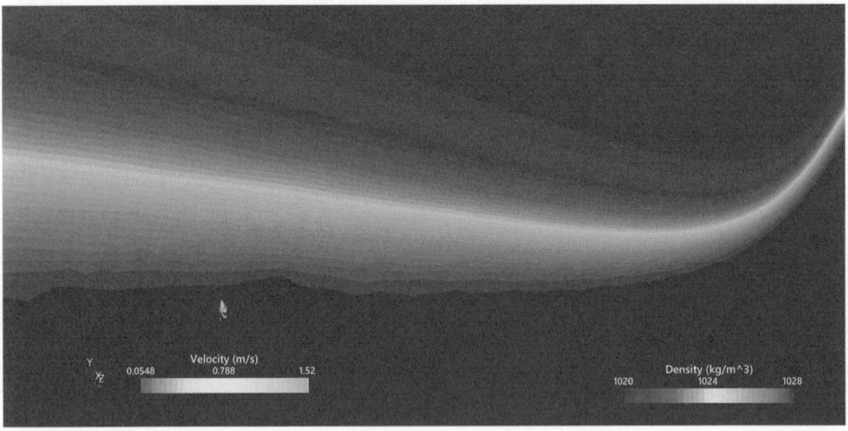

Fig. 6.17 Density contour at 69 s

figures reveal that the propeller is positioned in the middle of a continuous density stratification, with higher density near the propeller hub and lower density at the upper and lower blade regions. Additionally, a high-density zone forms upstream of the propeller, impeding the motion of the internal wave. Consequently, the propeller's thrust increases during this period, peaking at 0.237N—a 2.16% increase compared to the initial thrust (Figs. 6.18 and 6.19).

In the subsequent period (72–85 s), the propeller's thrust exhibits a significant downward trend, reaching its minimum value between 85 and 86 s at 0.18N—a 22.42% decrease compared to the initial thrust. This indicates that this phase is highly unfavorable for the propeller's propulsion efficiency. During this time, the propeller transitions from a high-density region to a low-density region, i.e., crossing

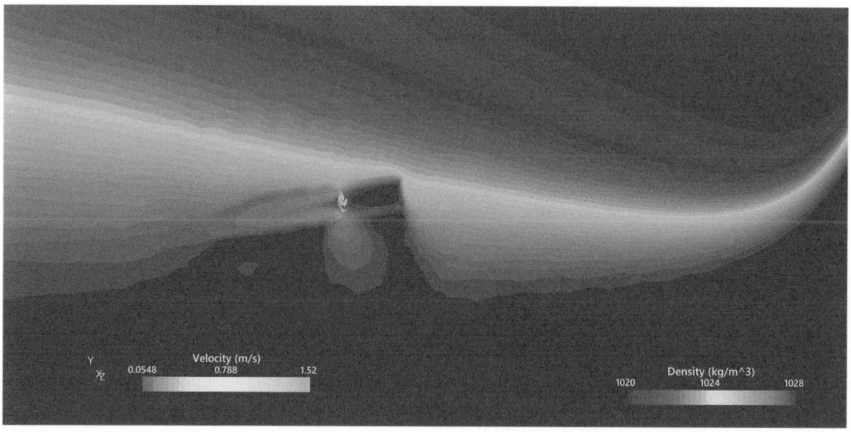

Fig. 6.18 Density contour at 72 s

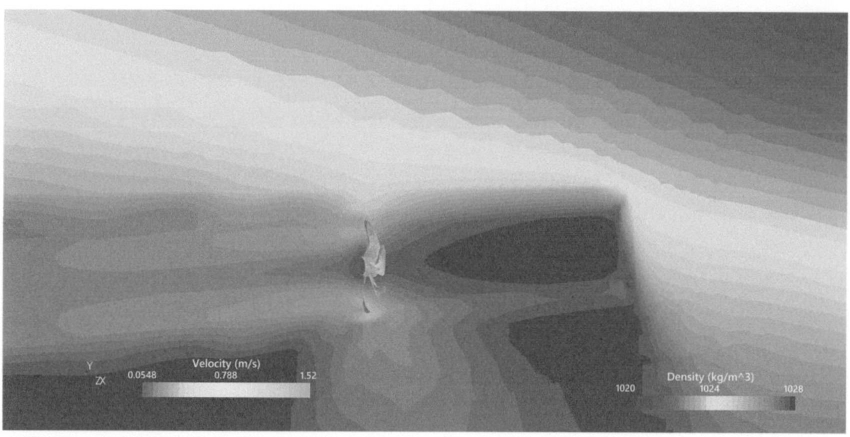

Fig. 6.19 Density contour around the propeller at 72 s

the leading surface of the internal solitary wave. By 75 s, the propeller has entered the low-density zone, but its rotational motion carries high-density fluid into the low-density zone and vice versa. The fluid density near the propeller hub remains much higher than that at the blade tips, leading to a noticeable slowdown in the thrust decline rate. By 77 s, the propeller fully enters the upper fluid layer, as shown in Figs. 6.20 and 6.21. Combined with the velocity field analysis, it is evident that the surrounding flow velocity continuously increases, causing the propeller's thrust to keep decreasing until it reaches its minimum at 85 s.

Starting from 86 s, the propeller's thrust begins to increase, reaching the highest value of the entire calculation process at 101 s, with a thrust of 0.248N, representing a 6.9% increase compared to the initial thrust. At 85 s, the propeller begins to traverse

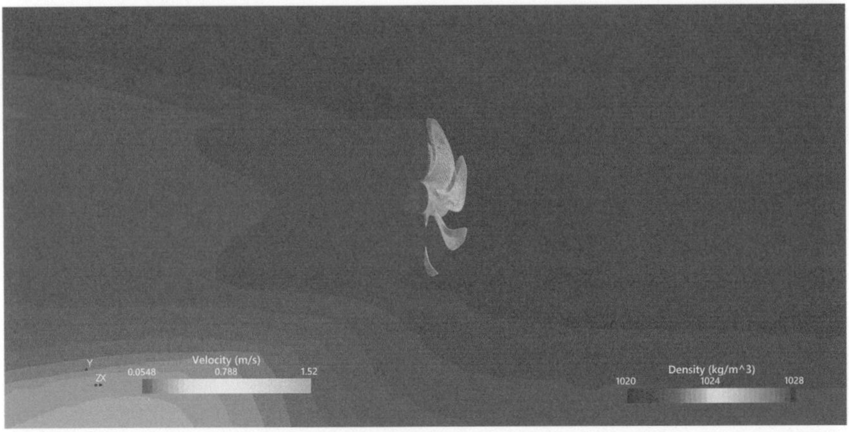

Fig. 6.20 Density contour around the propeller at 77 s

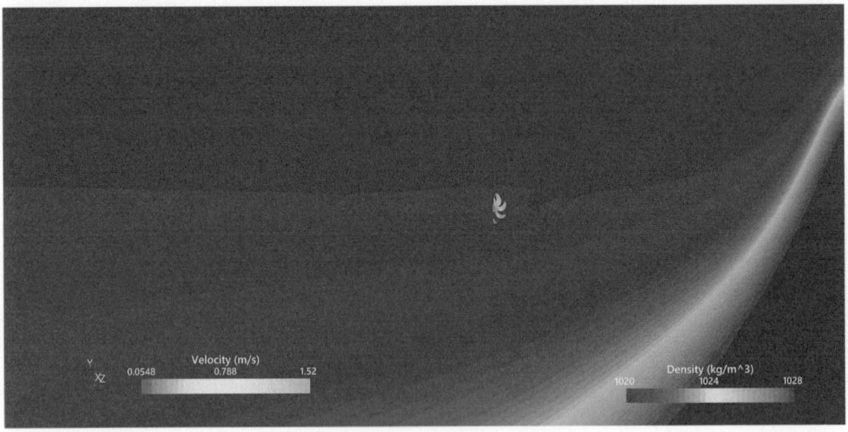

Fig. 6.21 Density contour around the propeller at 82 s

the rear wavefront of the internal solitary wave from the upper low-density fluid. During this time, the fluid velocity around the propeller gradually decreases, leading to an increase in thrust. The period from 86–101 s corresponds to the process of the internal wave's density transition layer passing through the propeller.

From 101–115 s, the propeller's thrust undergoes minor fluctuations. By analyzing the density and velocity contours, it is inferred that this is due to the propeller carrying out some low-density fluid after leaving the density transition layer, corresponding to the gradual detachment of low-density fluid from the propeller. After 115 s, as the internal wave moves away from the propeller, the thrust returns to its initial value and fluctuates slightly around it.

The propeller torque is closely related to thrust, exhibiting nearly identical trends. The reasons can be referenced from the thrust variation analysis.

Summarizing the above process reveals the primary influence of internal waves on the propeller: When the propeller crosses the internal wavefront, as the front wavefront approaches but has not yet reached the propeller, thrust and torque experience a slight decrease, with the magnitude dependent on the energy carried by the wave. As the front wavefront begins to traverse the propeller, thrust and torque rise slightly until the propeller reaches the middle of the density transition layer. Subsequently, thrust and torque undergo a significant drop—this phase requires extreme caution for the vessel. When the propeller is about to re-enter the density transition layer, thrust and torque reach their lowest values. Then, as the rear wavefront begins to traverse the propeller, these values start to rise again, peaking when the rear wavefront is about to leave the propeller. At this point, thrust and torque exceed their initial values. Finally, as the internal wave moves away, thrust and torque fluctuate and gradually return to their initial levels.

6.4.3 *Effect of Wave Amplitude on a Propeller Crossing an Internal Solitary Wave*

Based on the previous section, the amplitude of internal solitary waves was adjusted by increasing the wave energy, setting the amplitude to 1.71 m while keeping other parameters unchanged. This simulation investigated the interaction between internal solitary waves and propellers to explore the effects of different wave amplitudes. The thrust T and torque Q variation curves obtained from large-amplitude internal wave interactions were compared with those in Sect. 4.2, with results shown in Fig. 6.22.

The figure reveals that large-amplitude and small-amplitude internal waves affect propellers with fundamentally similar patterns, differing mainly in phase and amplitude. As energy increases with wave amplitude, wavelength gradually extends while wave speed continuously rises. Consequently, the maximum values of propeller thrust and torque under interaction also increase progressively, and the timing of thrust/torque curve variations advances accordingly. This occurs because higher-amplitude

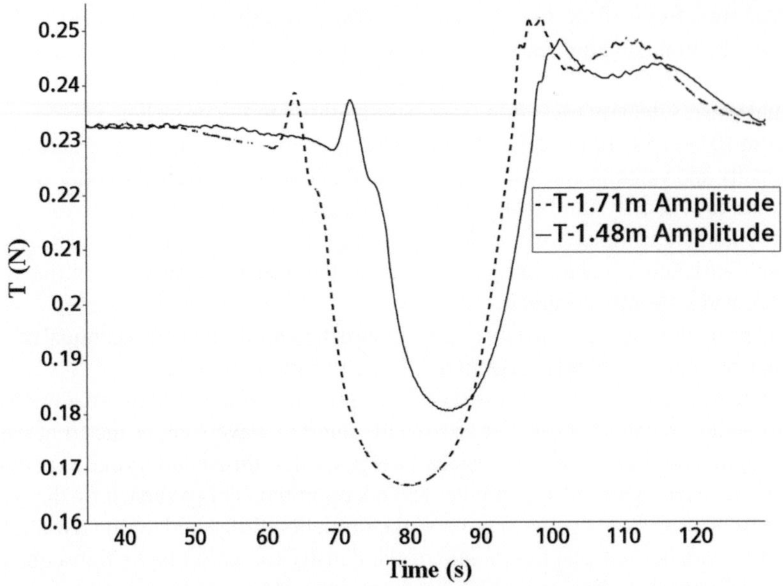

(a) thrust T variation curve

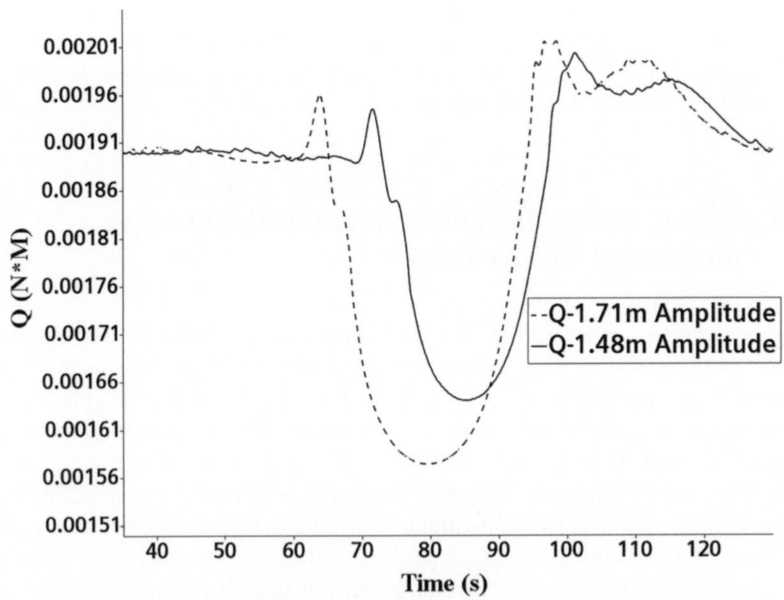

(b) torque Q variation curve

Fig. 6.22 Influence of internal solitary wave amplitude on propeller performance parameters internal solitary waves propagate faster, causing the propeller to encounter waves earlier, thus advancing the onset time of thrust/torque variations.

6.5 Summary

This chapter employed the SST $k - \omega$ turbulence model to establish an internal solitary wave numerical flume based on the fully nonlinear continuous stratification theory (DJL equations), investigating the effects of density-stratified structures and internal solitary waves on propeller hydrodynamic performance. Research indicates:

Continuous density stratification has minimal impact on propeller performance. When advance coefficient $J \leq 1.0$, the simulated thrust coefficient, torque coefficient, and open-water efficiency in stratified flow show excellent agreement with homogeneous flow results. Only when advance coefficient $J \geq 1.0$ do these parameters slightly exceed homogeneous flow values, with open-water efficiency showing the most significant variation.

Internal solitary waves substantially affect propeller performance parameters and wake fields. Notably, when the propeller traverses the wave front or remains above the wave surface (near wave trough), thrust and torque experience dramatic declines—up to 22.42% reduction from initial thrust, posing significant threats to vessel navigation safety.

Chapter 7
Influences of Internal Waves on Rudder

7.1 Introduction

As a crucial attitude control device for underwater vehicles, the rudder structure plays a vital role in the vehicle's motion state. Internal solitary waves may significantly impair the performance, efficiency, and wake field of the rudder structure, thereby posing a threat to the safe navigation of the vehicle. Investigating the influence of internal solitary waves on rudder structures also holds significant engineering value.

Based on the continuous stratified internal solitary wave numerical flume established in Chap. 2, this chapter first conducts a grid independence verification. Subsequently, numerical simulations are performed to obtain the dimensionless lift and drag coefficients of the rudder structure under uniform flow conditions at different rudder angles, and the results are compared with experimental data to validate the accuracy and reliability of the numerical method. On this basis, the interaction between internal solitary waves and the rudder structure is studied to explore the hydrodynamic performance changes of the rudder under internal wave effects, along with an analysis of the reasons behind these variations. Further, the position of the rudder structure, rudder angle, and the amplitude of the internal solitary wave are adjusted to investigate the influence of different working conditions on the performance and efficiency of the rudder, providing references for subsequent self-propulsion tests and steering strategies in practical engineering applications.

P. Du et al., *Internal Wave Hydrodynamics on Submersibles*,
https://doi.org/10.1007/978-981-95-4479-0_7

7.2 Computational Model and Validation

7.2.1 Geometric Model of the Rudder Structure

A rudder can be regarded as a low-aspect-ratio wing, and thus, parameters commonly used to describe wings are employed to characterize the geometric shape of the rudder structure. The main geometric elements include the following parameters:

(1) Rudder area AR: The projected area of the rudder blade profile fixed on the central longitudinal plane.
(2) Rudder height H: The vertical distance from the upper edge to the lower edge of the rudder blade along the rudder stock axis.
(3) Rudder width B: The horizontal distance between the leading and trailing edges of the rudder blade.
(4) Aspect ratio λ: The ratio of rudder height to rudder width. For a rectangular rudder, it is given by:

$$\lambda = \frac{H}{B} \tag{7.1}$$

The rudder structure selected for this study is the NACA0020 airfoil rudder, which is widely used in the design of underwater vehicles. Its main parameters are listed in Table 7.1, its airfoil cross-section is shown in Fig. 7.1, and its 3D model is illustrated in Fig. 7.2.

Table 7.1 Main parameters of the NACA0020 airfoil rudder

Parameter	Unit	Value
Rudder height H	m	0.24
Rudder width B	m	0.24
Rudder area A_R	m^2	0.0576
Aspect ratio λ	–	1

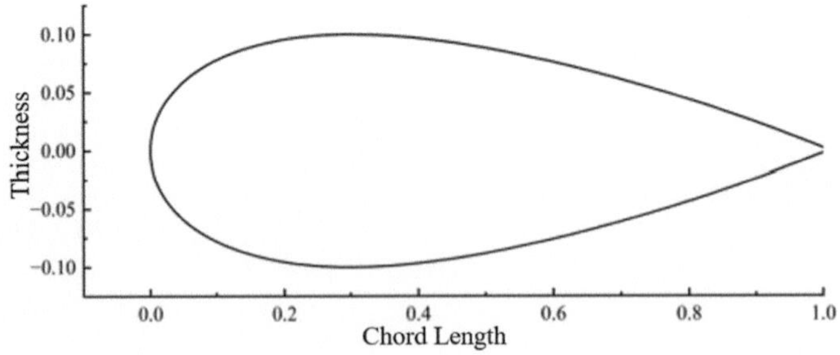

Fig. 7.1 Cross-section of the NACA airfoil

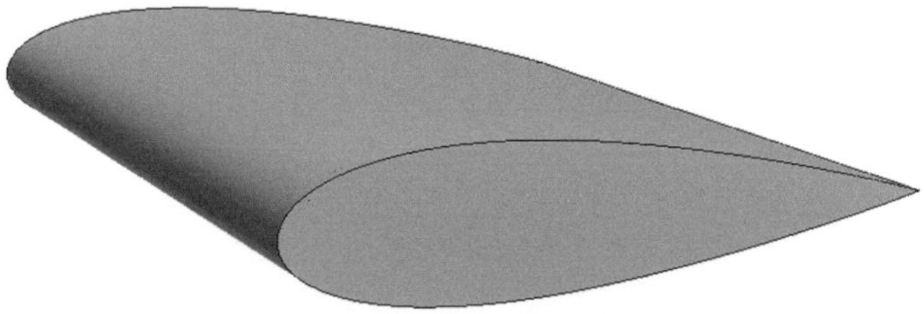

Fig. 7.2 3D model of the NACA airfoil rudder

7.2.2 Computational Domain and Grid Independence Verification

Before conducting numerical calculations of hydrodynamic performance, it is necessary to first perform grid independence verification of the model and validation of numerical methods to ensure computational accuracy. The computational domain for the validation case is a rectangular numerical water tank measuring 7 m × 4 m × 1.6 m, as shown in Fig. 7.3. The distance from the leading edge of the rudder structure to the inlet of the computational domain is 10 times the chord length, while the distance from the trailing edge to the outlet is 20 times the chord length. The distance from the rudder shaft center to both the upper and lower walls of the computational domain is 8 times the chord length, and the distance to the front and rear walls is 4 times the chord length. The inlet of the computational domain is set as a velocity inlet, the outlet as a pressure outlet, and all other boundaries are set as no-slip walls.

Before conducting numerical simulations, the meshing of the rudder structure and computational domain is crucial, as the quality of the grid directly affects the precision and accuracy of the calculation results. In this chapter's study, a cut-cell mesh generator is employed to discretize the research object and computational domain. The cut-cell mesh combines the advantages of structured and unstructured grids. The

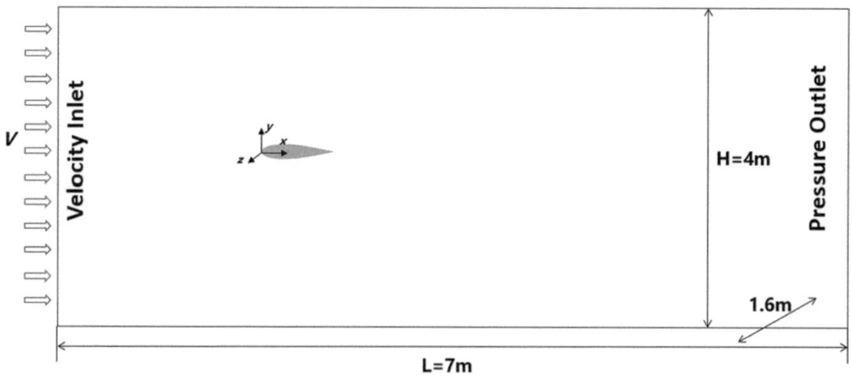

Fig. 7.3 Computational domain for validation case

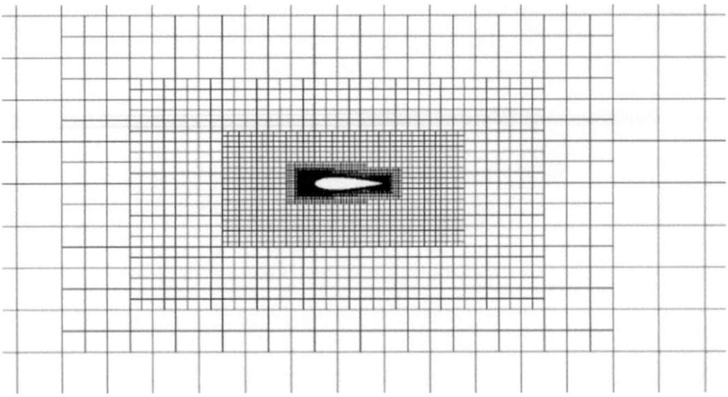

Fig. 7.4 Computational domain meshing

flow characteristics and hydrodynamic features around the rudder structure are the focus of this chapter, so mesh refinement is required around the rudder's leading edge, trailing edge, and head. Meanwhile, to improve computational efficiency, the grid size is appropriately enlarged in the far field, which is farther from the rudder structure. The boundary layer can be divided into the viscous sublayer, buffer layer, and log-law layer based on the distance along the wall-normal direction. Among them, the flow in the log-law layer is fully developed turbulent flow, and the velocity distribution follows the logarithmic law. To better study the interaction between internal solitary waves and the rudder structure in subsequent research, fully developed turbulence must be considered. Therefore, the first layer of grid control points must be located in the log-law region of the boundary layer. The boundary layer region can be determined by the dimensionless y+ value. The number of boundary layers for the rudder structure is set to 25, and the overall wall's y+ value is around 1. The minimum grid size is 1×10^{-4} m, and the grid growth rate is set to 1.2. The meshing of the computational domain and rudder structure is shown in Figs. 7.4 and 7.5.

To avoid the dependence of numerical simulation results on grid precision, three sets of grids were divided to verify the grid independence of the rudder structure. According to GCI (Grid Convergence the Index) method [79], the basic grid size was proportionally refined with a growth rate of. During the refinement process, the grid was globally encrypted to verify grid independence in all directions, while ensuring that the scale of the first layer of wall grid remained within a reasonable range. The overall quantities of the three sets of grids, from coarse to fine, are shown in Table 7.2.

The fluid density was set to 997.561 kg/m^3, the inflow velocity V = 3 m/s, and the rudder angle was 10°. The SST turbulence model was selected. The k–ω turbulence model is adopted, with the PISO algorithm used for numerical solution. The pressure gradient at discrete nodes is calculated using the least squares method, the pressure interpolation scheme employs volume force weighting, and first-order implicit time discretization is applied. The momentum equation, convection term, and turbulence transport term are set to second-order upwind scheme, with a time step of 0.01 s. The

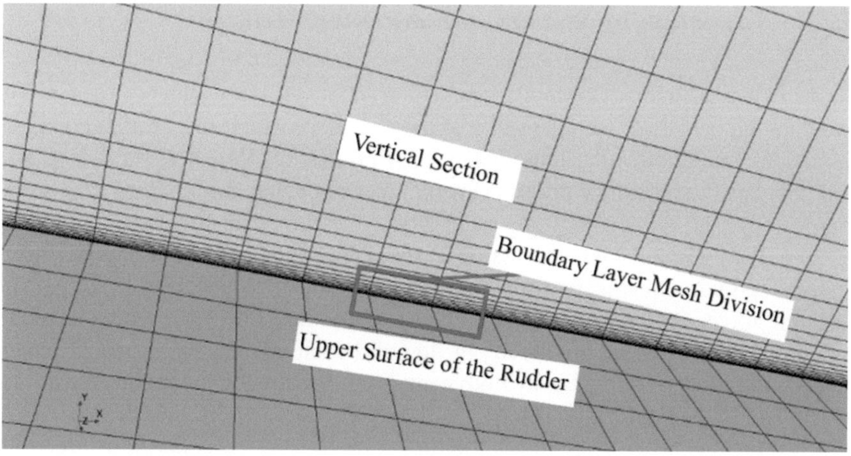

Fig. 7.5 Boundary layer grid

Table 7.2 Grid independence verification

Grid name	Coarse	Medium	Fine
Total grid count	1,023,668	2,807,513	4,321,748

lift coefficient (CL) and drag coefficient (CD) of the rudder structure are monitored, and the calculation results for the three meshes are shown in Table 7.3.

As can be seen from Table 7.3, the lift and drag coefficients of the rudder structure have basically stabilized under medium-density mesh, and the calculation results no longer change significantly with further mesh refinement. The errors between the lift coefficients obtained from all three meshes and the experimental values are slightly larger, while the errors in drag coefficients are smaller. The errors between the lift coefficients calculated by medium and fine meshes and the experimental values are 8.95% and 8.35% respectively, with only a 0.6% difference. The errors between the calculated drag coefficients and experimental values are 6.74% and 6.29% respectively, with only a 0.45% difference. Therefore, it can be concluded that medium-density mesh can already accurately calculate the performance parameters of the rudder structure, and subsequent simulations will also use medium-density mesh for calculations.

Table 7.3 Calculation results of grid independence verification case

Grid name	Coarse	Medium	Fine	Experimental value
Lift coefficient C_L	0.3802	0.3650	0.3630	0.3350
Drag coefficient C_D	0.0480	0.0475	0.0473	0.0445
C_L error (%)	13.49	8.95	8.35	–
C_D error (%)	7.86	6.74	6.29	–

7.2.3 Validation of Numerical Simulation Method

To verify that the numerical simulation method adopted in this chapter can accurately simulate the performance parameters of the rudder structure, this section conducts numerical simulation on the hydrodynamic performance of NACA0020 airfoil rudder and compares the results with experimental measurements from towing tank tests, thereby ensuring the accuracy and reliability of the numerical simulation method.

First, the hydrodynamic expression of the open-water rudder needs to be introduced. The hydrodynamic performance of an open-water rudder refers to the hydrodynamic forces and moments acting on the rudder in a flow field where only the rudder exists. Assuming the incoming flow direction is along the angle of attack towards the rudder profile, the hydrodynamic forces acting on the rudder are shown in Fig. 7.6. In the figure, α represents the rudder's angle of attack, which is the angle between the chord line direction of the rudder profile and the water flow direction, generally referred to as the rudder angle; a represents the distance from the rudder stock axis to the leading edge, b represents the chord length of the rudder profile; represents the pressure center of the rudder, referring to the distance from the action point of the resultant force to the leading edge of the rudder; P represents the resultant force acting on the rudder, with P_y being the lift force perpendicular to the incoming flow direction, and P_x being the drag force parallel to the incoming flow direction.

To enable unified comparison of the rudder's hydrodynamic performance parameters, the following dimensionless coefficients are commonly used in engineering practice to characterize the hydrodynamic performance of the rudder structure:

Lift coefficient:

$$C_L = \frac{P_y}{\frac{1}{2}\rho V^2 A_R} \tag{7.2}$$

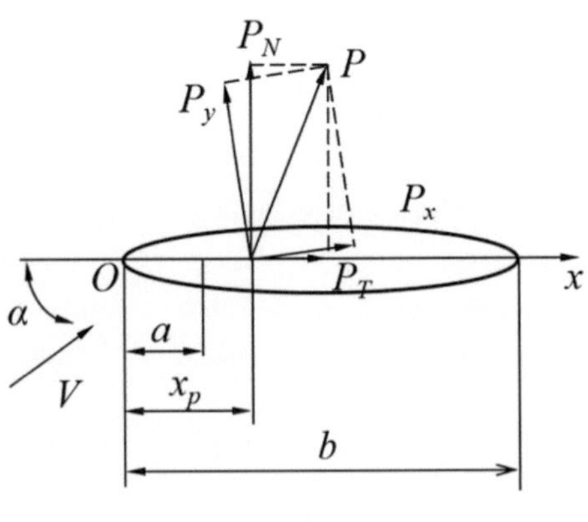

Fig. 7.6 Schematic diagram of forces acting on an open-water rudder

Drag coefficient:

$$C_D = \frac{P_x}{\frac{1}{2}\rho V^2 A_R} \tag{7.3}$$

Moment coefficient:

$$C_M = \frac{M}{\frac{1}{2}\rho V^2 A_R b} \tag{7.4}$$

where V represents the incoming flow velocity; A_R represents the rudder area.

Generally speaking, the lift coefficient and drag coefficient of the rudder structure can already well reflect the hydrodynamic forces acting on the rudder structure. Therefore, this paper mainly focuses on the hydrodynamic performance parameters of the rudder, namely the lift coefficient C_L and drag coefficient C_D.

In the validation of numerical simulation methods, the numerical water tank established in Sect. 7.2.2 was used. The mesh division adopted medium-density grid settings, and the numerical calculation method remained consistent with that in Sect. 7.2.2. On this basis, the rudder angle of the rudder structure was changed, and the lift coefficient C_L and drag coefficient C_D of the rudder structure at different rudder angles were monitored. A total of six working conditions with different rudder angles were set, and the specific case settings are shown in Table 7.4.

The comparison between the open-water rudder lift coefficient, drag coefficient obtained from numerical simulation and experimental results is shown in Table 7.5 and Fig. 7.7.

From Table 7.5, it can be found that the results obtained from numerical simulation are basically consistent with the experimental results. The error between the calculated lift coefficient and the experimental results is larger than that of the drag coefficient, but the maximum error does not exceed 10%. It can be considered that the numerical simulation method adopted in this chapter can well reflect the real hydrodynamic performance of the rudder structure. The same numerical simulation method will also be used in subsequent simulations.

From the variation curves of lift coefficient and drag coefficient shown in Fig. 7.7, it can be found that for the NACA0020 airfoil rudder selected in this chapter, both the

Table 7.4 Numerical method validation case settings

Case setting	Rudder angle α	Fluid density ρ (kg/m^3)	Incoming flow velocity V (m/s)
Case 0–1	10°	997.561	3
Case 0–2	15°		
Case 0–3	20°		
Case 0–4	25°		
Case 0–5	30°		
Case 0–6	35°		

Table 7.5 Comparison between CFD calculation results and experimental results

Rudder angle α	Lift coefficient C_L			Drag coefficient C_D		
	CFD	Experimental value	Error (%)	CFD	Experimental value	Error (%)
10°	0.365	0.335	8.95	0.0480	0.0445	7.86
15°	0.549	0.502	9.36	0.0920	0.0900	2.22
20°	0.712	0.673	5.79	0.1546	0.1556	0.64
25°	0.836	0.833	0.36	0.2242	0.2394	6.34
30°	1.044	1.020	2.35	0.3450	0.3490	1.14
35°	0.535	–	–	0.3860	–	–

lift coefficient and drag coefficient continuously increase as the rudder angle changes from 0° to 30°. From the velocity vector diagrams of different rudder angle conditions shown in Fig. 7.8, it can be found that the fluid velocity above the leading edge of the rudder continuously increases, the disturbance to the entire uniform flow field also continuously strengthens, and the zero-velocity region near the rudder blade surface on the leeward side of the rudder structure continuously expands.

When the rudder angle reaches 35°, the lift coefficient C_L rapidly decreases to 0.535, which is 51.2% of the lift coefficient at 30° rudder angle. At this point, the rudder structure can be considered to have stalled, with flow separation occurring in the flow field. As shown in Fig. 7.9a, the flow velocity on the entire backflow surface of the rudder blade approaches zero, and periodic vortex shedding occurs. This results in periodic high-pressure zones on the backflow surface of the rudder blade, as shown in Fig. 7.9b, leading to a sharp drop in the lift coefficient of the rudder. Therefore, it can be concluded that the NACA0020 airfoil rudder's stall angle is approximately 35°.

7.3 Study on Hydrodynamic Characteristics of Rudder Under Internal Solitary Wave Action

Based on the validation of numerical simulation methods in the previous section, this section investigates the interaction between internal solitary waves and rudder structures in a continuously stratified internal solitary wave flume. By adjusting parameters such as the rudder's submerged depth, rudder angle, and wave amplitude of internal solitary waves, the influence patterns of internal solitary waves on the hydrodynamic performance of rudder structures under different working conditions are explored, providing valuable references for steering strategies of underwater vehicles in real marine environments.

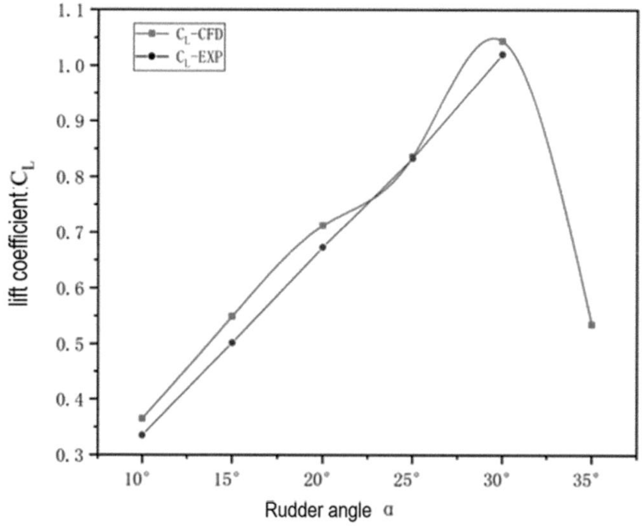

(a)Lift coefficient

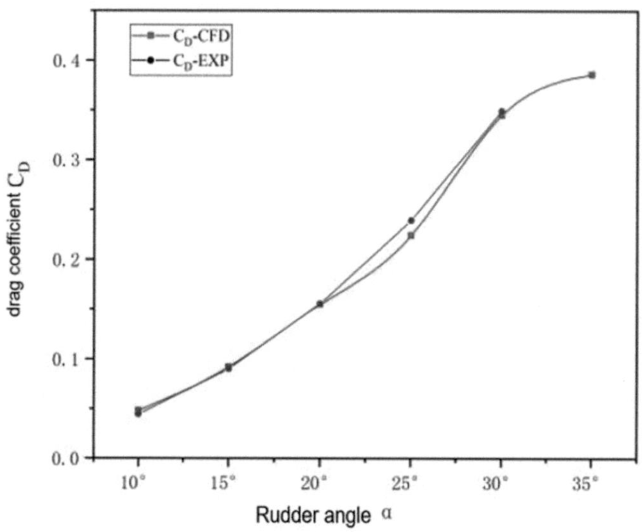

(b) Drag coefficient

Fig. 7.7 Comparison between CFD calculation results and experimental results

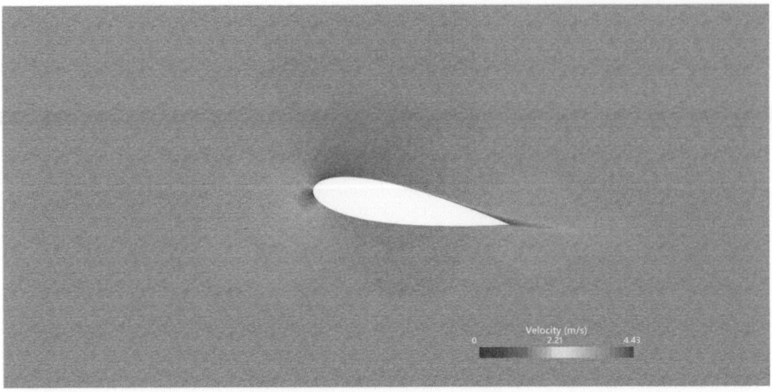

(a) 10º rudder angle

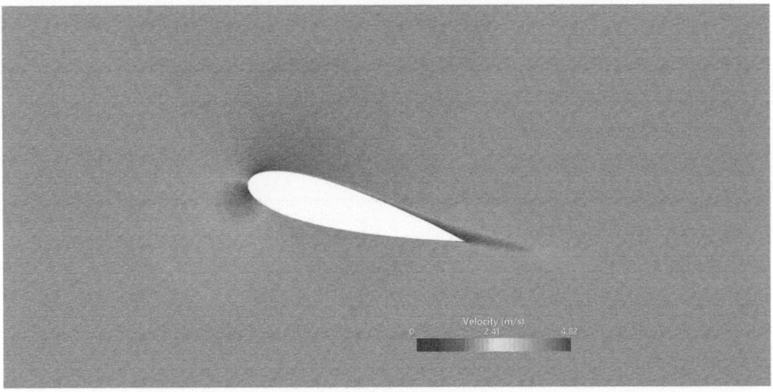

(b) 15º rudder angle

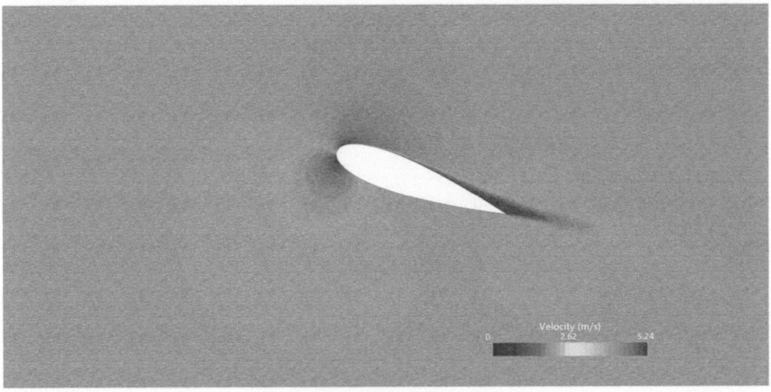

(c) 20º rudder angle

Fig. 7.8 Velocity vector diagrams under different rudder angle conditions

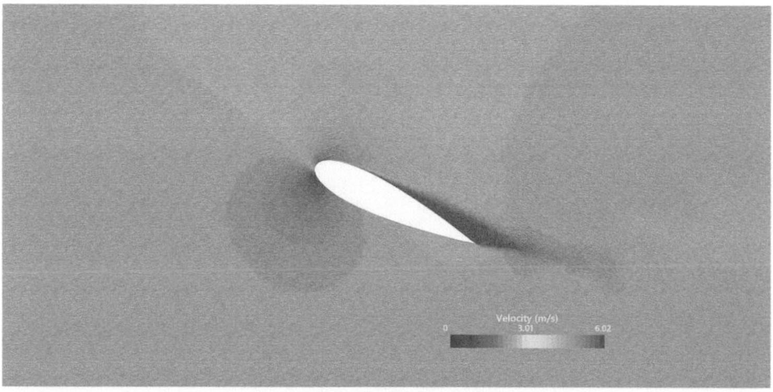

(d) 25° rudder angle

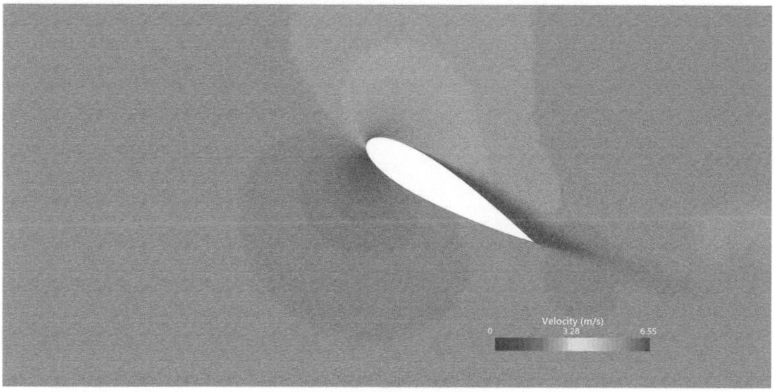

(e) 30° rudder angle

Fig. 7.8 (continued)

7.3.1 Model and Computational Domain Setup

This section establishes a numerical model for the interaction between continuously stratified internal solitary waves and rudder structures, based on the internal solitary wave numerical flume established in Sect. 3.3 and the validated rudder blade grid and numerical simulation methods in Sect. 7.2. To ensure that the scale of internal solitary waves relative to the rudder structure matches real-world conditions, the NACA0020 airfoil rudder model is scaled down by a factor of 0.5, with specific parameters listed in Table 7.6.

Meanwhile, to provide stable propagation space for internal solitary waves and ensure the integrity and stability of the waveform when encountering the rudder, the computational domain from Sect. 7.2 is expanded while keeping other conditions unchanged. The numerical flume computational domain model used in this section is shown in Fig. 7.10. The numerical flume is 70 m long and 5 m high, divided into three main regions: the wave-making interpolation zone, the interaction zone, and

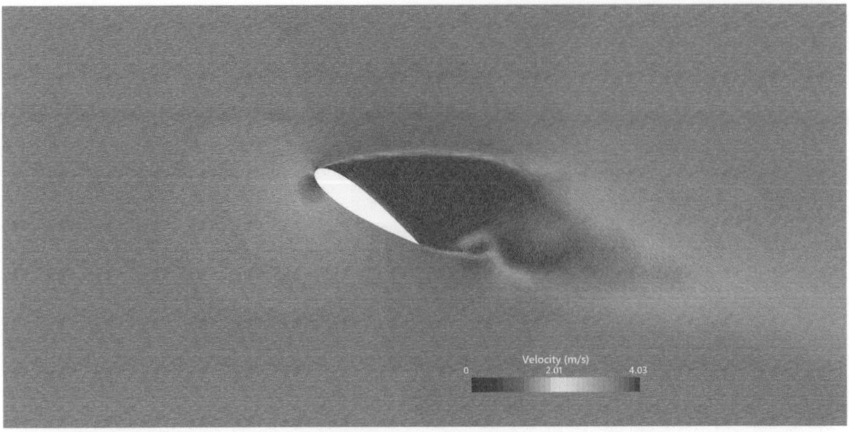

(a) Velocity vector diagram

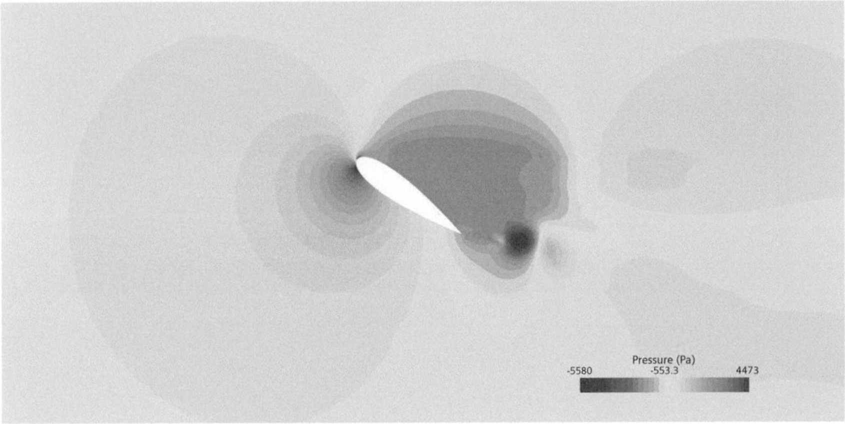

(b) Pressure contour

Fig. 7.9 Flow field under 35° rudder angle condition

Table 7.6 Main parameters of scaled rudder model

Parameter	Unit	Value
Rudder height H	m	0.12
Rudder width B	m	0.12
Rudder area A_R	m^2	0.0144
Aspect ratio λ	λ	−1
Maximum vertical height h_0	m	0.02

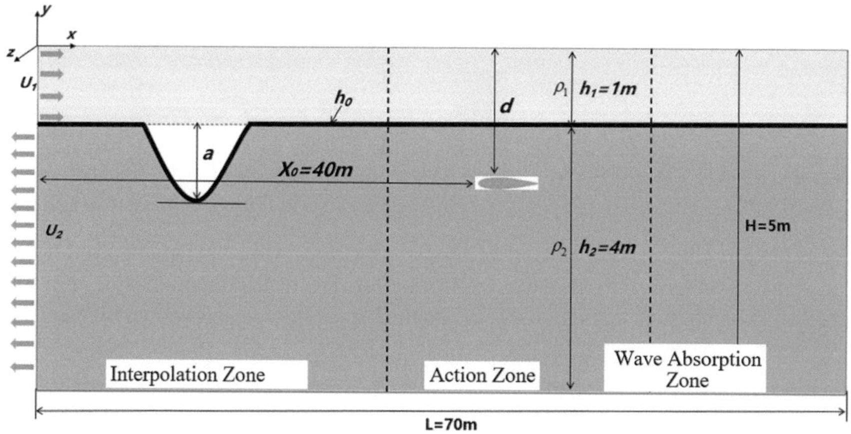

Fig. 7.10 Computational domain model for interaction between internal solitary waves and rudder

the wave-absorbing zone. The grid division method for the computational domain remains consistent with that described in Sect. 2.5. The rudder is located 40 m from the flume entrance.

This chapter still simulates continuous internal solitary waves based on the DJL equation. Considering the thickness of the density pycnocline, the liquid in the tank can be regarded as three layers. The thickness of the upper fluid is $h_1 = 1.0$, with a density of $\rho_1 = 1020\,\text{kg/m}^3$. The thickness of the lower fluid is $h_2 = 1.0$, with a density of $\rho_2 = 1028\,\text{kg/m}^3$. The thickness of the continuous density pycnocline is $h_0 = 1.0$, where the density increases gradually from top to bottom, with a density variation range of $1020-1028\,\text{kg/m}^3$. The dynamic viscosity of the fluid is $\mu = 0.001\,\text{Ns/m}^2$. The vertical plane at the entrance of the numerical tank is defined as $x = 0$, and the upper wall of the tank is $y = 0$. The origin of the coordinate system is located at the upper-left corner of the numerical tank, as shown in the coordinate system position in Fig. 7.10. The direction of internal solitary wave propagation is defined as the positive direction of the x-axis, and the vertically upward direction is the positive direction of the y-axis. A positive rudder angle is formed by clockwise rotation of the rudder blade, and the vertical distance between the rudder's center of gravity and the upper surface of the numerical tank is defined as the submerged depth d.

The bottom of the tank is set as a wall (Wall), while the remaining boundary conditions are set as symmetry boundaries (Symmetry). The numerical solver selected is an unsteady transient solver, and the finite volume method (Finite Volume Method, FVM) is used to discretize the governing equations. The Mixture multiphase flow model is employed to simulate the multiphase fluid. The turbulence model selected is SST $k-\omega$, and the numerical solution method adopted the PISO algorithm. The pressure gradient at discrete nodes was calculated using the least squares method, with the pressure interpolation scheme selected as body-force-weighted. The time

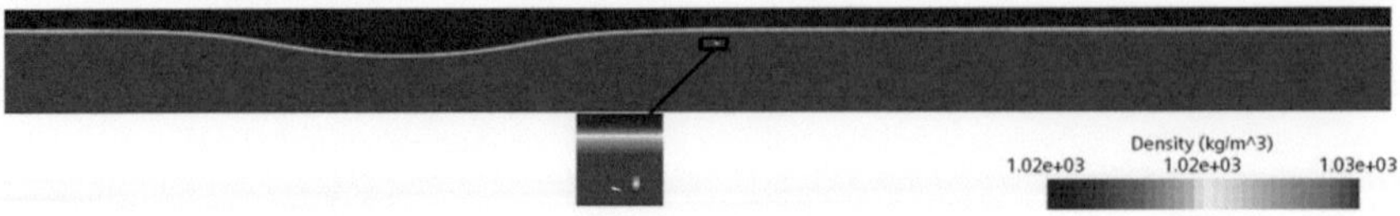

Fig. 7.11 Initialized flow field of the interaction between internal solitary waves and the rudder

discretization employed a first-order implicit method, while the momentum equation, convection term, and turbulence transport term were set to second-order upwind scheme. The time step was set to 0.05 s. The initialized flow field is shown in Fig. 7.11.

7.3.2 Influence of Submergence Depth

First, the influence of different positions relative to the internal solitary wave on the hydrodynamic performance of the rudder structure was investigated. During the simulation, parameters such as the amplitude and speed of the internal solitary wave were kept consistent. In this section, the internal solitary wave amplitude WA = 1.255 m, wave speed V = 0.3 m/s, and wavelength WL = 25.961 m. The rudder angle remained consistent for rudder structures at different positions, while the submergence depth of the rudder structure was adjusted. When the submergence depth $d < -1$ m, the rudder structure was located above the wave surface of the internal solitary wave; when $-1 \text{ m} < d < -2.255$ m, the rudder structure penetrated through the wave surface of the internal solitary wave; when $d > -2.255$ m, the rudder structure was located below the wave trough of the internal solitary wave. This section established a total of 6 working conditions (Cases 1–6), with specific condition parameters shown in Table 7.7. Among them, Cases 1–3 compared the hydrodynamic performance of rudder structures with 0° rudder angle at different positions, while Cases 4–6 compared those with 10° rudder angle at different positions. Due to the mixing of different phase fluids in continuous stratified internal solitary waves and the relatively complex induced velocity field, this study directly monitored the changes in lift and resistance forces on the rudder structure.

(1) Interaction between internal solitary wave and 0° rudder structure at different submergence depths.

The investigation of 0° rudder angle case aims to provide reference for the force situation on rudder structures when underwater vehicles encounter internal waves without maneuvering in real environments. Figure 7.12 shows the numerical calculation results of lift and resistance forces on the 0° rudder structure at different positions relative to the internal solitary wave.

Figure 7.12a presents the lift variation curves of the rudder structure with a 0° rudder angle under the influence of internal solitary waves at different submerged depths. From the figure, it can be observed that the rudder structure located above

Table 7.7 Working condition settings for rudders with different submergence depths

Case	Submergence depth d' (m)	State	Rudder angle α	Wave Amplitude a (m)
Case 1	−0.5	Above wave surface	0°	1.225
Case 2	−1.5	Crossing wave surface		
Case 3	−3.0	Below wave surface		
Case 4	−0.5	Above wave surface	10°	
Case 5	−1.5	Crossing wave surface		
Case 6	−3.0	Below wave surface		

the wave surface, i.e., in the upper fluid layer (Case 1), is less affected by the internal solitary wave. Only when the right wave surface gradually moves beneath the rudder structure does a slight decrease occur, subjecting it to a downward vertical force. The reason for this is that the velocity field induced by the internal solitary wave rotates clockwise. As the rudder structure approaches the right wave surface, the flow field exhibits a downward vertical velocity. As the wave surface moves rightward, the downward vertical velocity gradually decreases, and the lift force gradually increases.

The rudder structure crossing the internal solitary wave surface (Case 2) exhibits a lift trend similar to that of Case 1 before 80 s, differing only in amplitude. The downward vertical force experienced by the rudder in Case 2 is significantly greater than in Case 1. This is because the initial position of the rudder in Case 2 is 0.5 m below the wave surface, where the downward vertical velocity is the largest in the flow field. After the right wave surface passes the rudder structure, the rudder enters the upper fluid layer, and its force trend becomes essentially the same as in Case 1, oscillating slightly around zero.

After 80 s, when the rudder structure crosses the left wave surface of the internal solitary wave, the lift variation trend in Case 2 differs significantly from that in Case 1. At this stage, the rudder structure experiences an upward lift force because the flow field in this region has an upward velocity. As the rudder structure traverses the internal solitary wave surface, the upward velocity gradually increases, leading to a corresponding rise in lift. Around 89 s, the lift reaches its maximum value, at which point the rudder structure is about to transition from the upper fluid layer into the continuous density transition layer. The surrounding velocity field at this moment is illustrated in Fig. 7.13a.

When the rudder structure is located below the trough of the internal wave (Case 3), the variation in its vertical force exhibits severe oscillations compared to the other two positions, with the amplitude of the force being significantly larger. However, the overall trend of variation is relatively consistent with Case 2. The difference

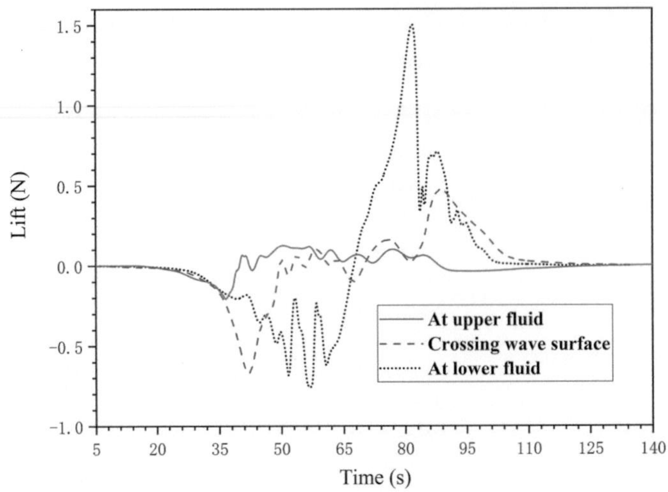

(a) Lift variation

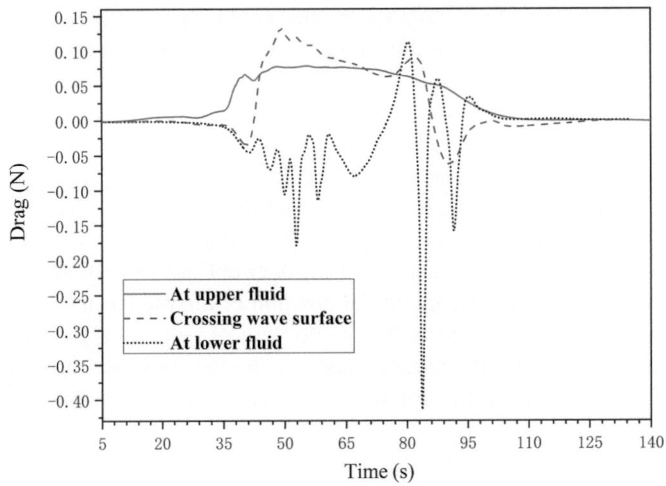

(b) Drag variation

Fig. 7.12 Interaction between internal solitary wave and rudder structures with different submergence depths at 0° rudder angle

lies in the fact that the rudder structure in Case 3 remains entirely within the lower fluid layer, so there is no stage where the lift force oscillates slightly around zero. When the rudder structure beneath the wave surface approaches the right wavefront of the internal solitary wave, the vertical force it experiences is still caused by the clockwise motion of the fluid. Between 47 and 65 s, the lift force on the rudder structure undergoes intense oscillations. During this stage, the rudder structure is

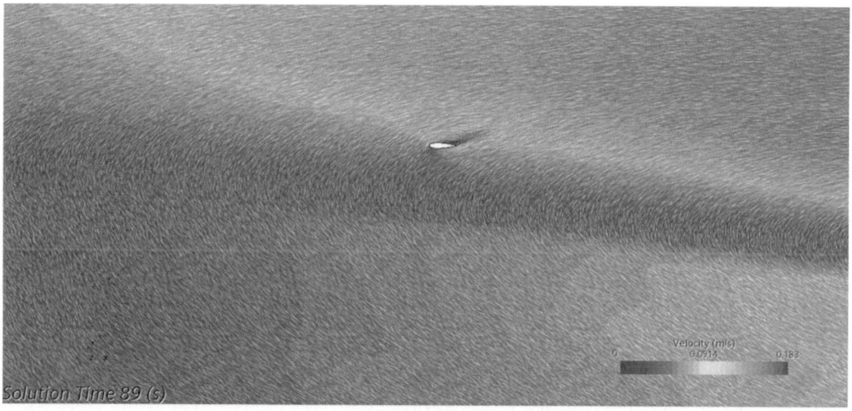

(a) Velocity vector field

(b) Density distribution diagram

Fig. 7.13 Flow field when rudder structure traverses the left wave surface of internal solitary wave

located below the right wavefront of the internal wave, as shown in Fig. 7.14a. The surrounding flow field exhibits downward-leftward fluid motion, as illustrated in Fig. 7.14b. The rudder structure continuously experiences a downward vertical force until it is directly below the wave trough. Analysis of the velocity vector field during this phase reveals that the oscillatory behavior arises because the fluid flow direction is nearly perpendicular to the rudder structure. The upper surface of the rudder blade exerts significant resistance on the fluid, while the flow velocity across the entire lower surface approaches zero. Additionally, periodic vortex shedding occurs, leading to the periodic formation of high-pressure zones on the leeward side of the rudder blade and flow separation. This stage poses a considerable threat to the performance and safety of the rudder structure. After the trough of the internal solitary wave passes the rudder structure, the rudder is positioned below the left wavefront of the internal solitary wave. During this phase, the surrounding flow field moves upward-leftward, resulting in an upward vertical force on the rudder. As the internal solitary wave moves rightward, the upward velocity gradually increases, causing the lift force to continuously rise until it reaches its maximum around 81 s. At this point, high-speed zones and vortex structures appear on the upper surface of the rudder, while the pressure on the lower surface exceeds that on the upper surface, as shown in Fig. 7.15.

Figure 7.12b presents the variation curve of drag force on the rudder structure at 0° angle under the influence of internal solitary waves at different submergence depths. Compared to lift force, the drag has a relatively minor impact on the performance of the rudder structure. The primary reason for drag variation lies in the direction and magnitude of flow velocity in the field. For the rudder structure above the wave surface (Case 1), as the internal solitary wave approaches, the upper-layer fluid velocity is mainly along the positive X-axis direction, resulting in a drag force consistently greater than zero. As the internal solitary wave moves rightward, the horizontal velocity increases, leading to a corresponding rise in drag. The rudder structure crossing the wave surface (Case 2) initially resides in the lower-layer fluid, where the flow direction is along the negative X-axis, causing a negative horizontal force. Subsequently, upon passing through the right wave surface and entering the upper-layer fluid, the surrounding flow velocity shifts to the positive X-axis direction, turning the drag from negative to positive. Compared to Case 1, the rudder in Case 2 is closer to the wave surface, experiencing faster flow speeds, thus exhibiting larger drag amplitudes. Eventually, as the rudder passes through the left wave surface and re-enters the lower-layer fluid, it again encounters drag in the opposite direction. The rudder structure below the wave surface (Case 3) remains submerged in the lower-layer fluid, primarily subjected to drag along the negative X-axis. Due to flow separation as previously described, the drag variation exhibits intense oscillations. Notably, between 80 and 95 s, the drag direction reverses with significant amplitude fluctuations. Analysis of the velocity field during this phase reveals periodic vortex structures forming on the upper surface of the rudder. These vortices cause localized flow velocities to shift toward the positive X-axis, altering drag direction. As vortices shed, the flow velocity restores, rapidly reducing drag and reverting its direction. This process is illustrated in Fig. 7.16.

(2) Interaction between internal solitary waves and rudder structures at 10° angle under different submergence depths.

Investigating the 10° rudder angle scenario aims to provide reference for real-world underwater vehicles encountering internal waves during maneuvering. The presence of a rudder angle increases the lift force on the rudder itself. Figure 7.17 displays the numerically calculated lift and drag forces on the 10° rudder structure at different positions relative to the internal solitary wave.

Figure 7.17a displays the lift variation curves of the rudder structure with 10° rudder angle at different submergence depths under the action of internal solitary waves. The rudder structure located in the upper fluid layer (Case 4) experiences minimal lift variation. Due to the presence of rudder angle, when the right wave

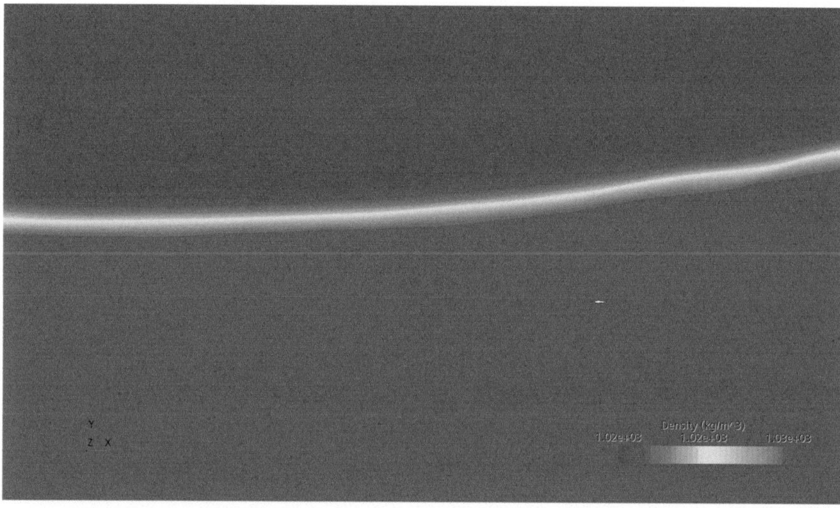

(a) Position of the rudder structure relative to the wave surface

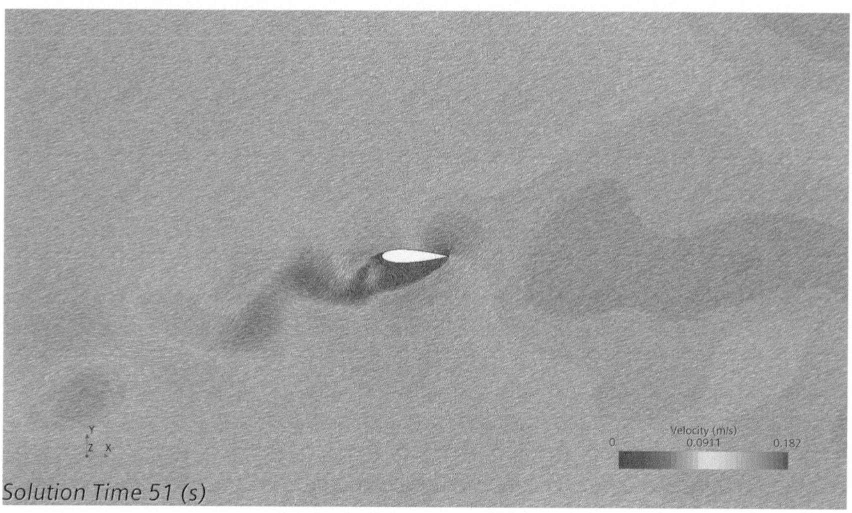

(b) Velocity vector field

Fig. 7.14 Flow field when the rudder structure is located at the right wavefront of the internal solitary wave

surface of the internal solitary wave approaches, the angle between the rudder's cross-section and the direction of the velocity field induced by the internal solitary wave decreases. Therefore, when above the right wave surface, almost no downward vertical force is generated on the rudder structure, and the lift fluctuates between 0.1 and 0.2 N. For the rudder structure crossing the internal solitary wave surface

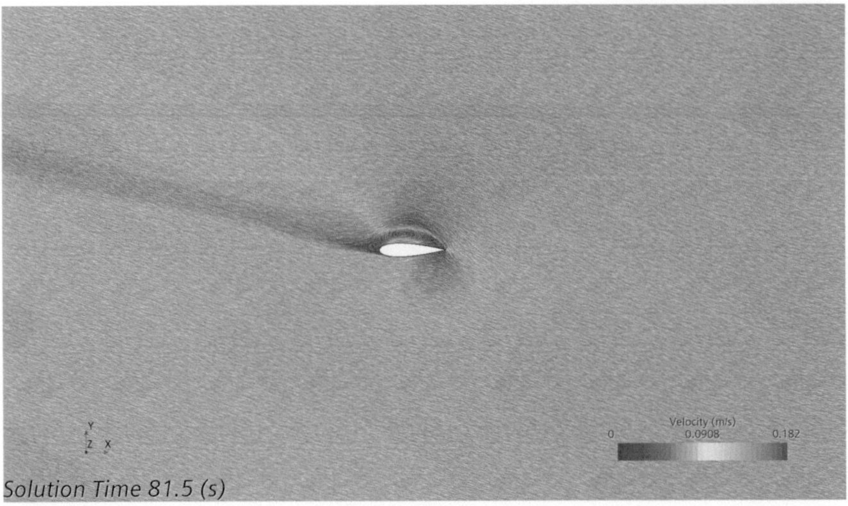

(a) Velocity vector field

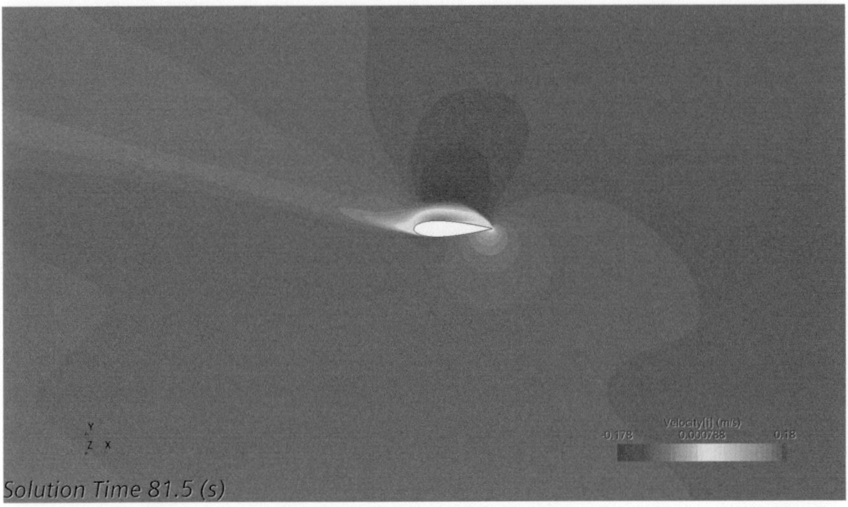

(b) Horizontal velocity field

Fig. 7.15 Flow field when the rudder structure is located at the left wavefront of the internal solitary wave

(Case 5), as it initially resides in the lower fluid layer, it is subjected to downward vertical force. After entering the upper fluid layer, due to the rudder angle, the rudder structure generates lift. As the horizontal velocity in the positive X-direction gradually increases, the lift also increases until the rudder is positioned above the trough of the internal solitary wave, where the horizontal velocity reaches its maximum.

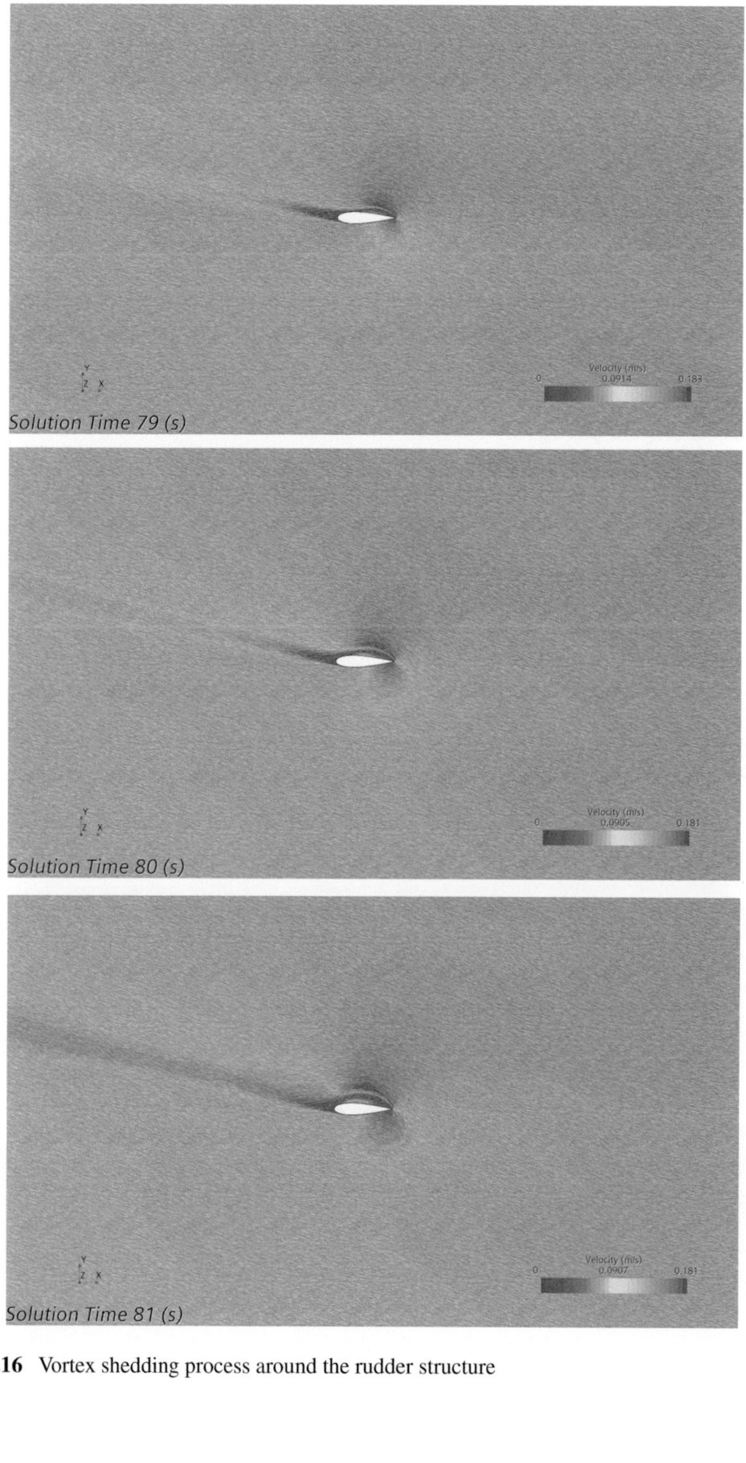

Fig. 7.16 Vortex shedding process around the rudder structure

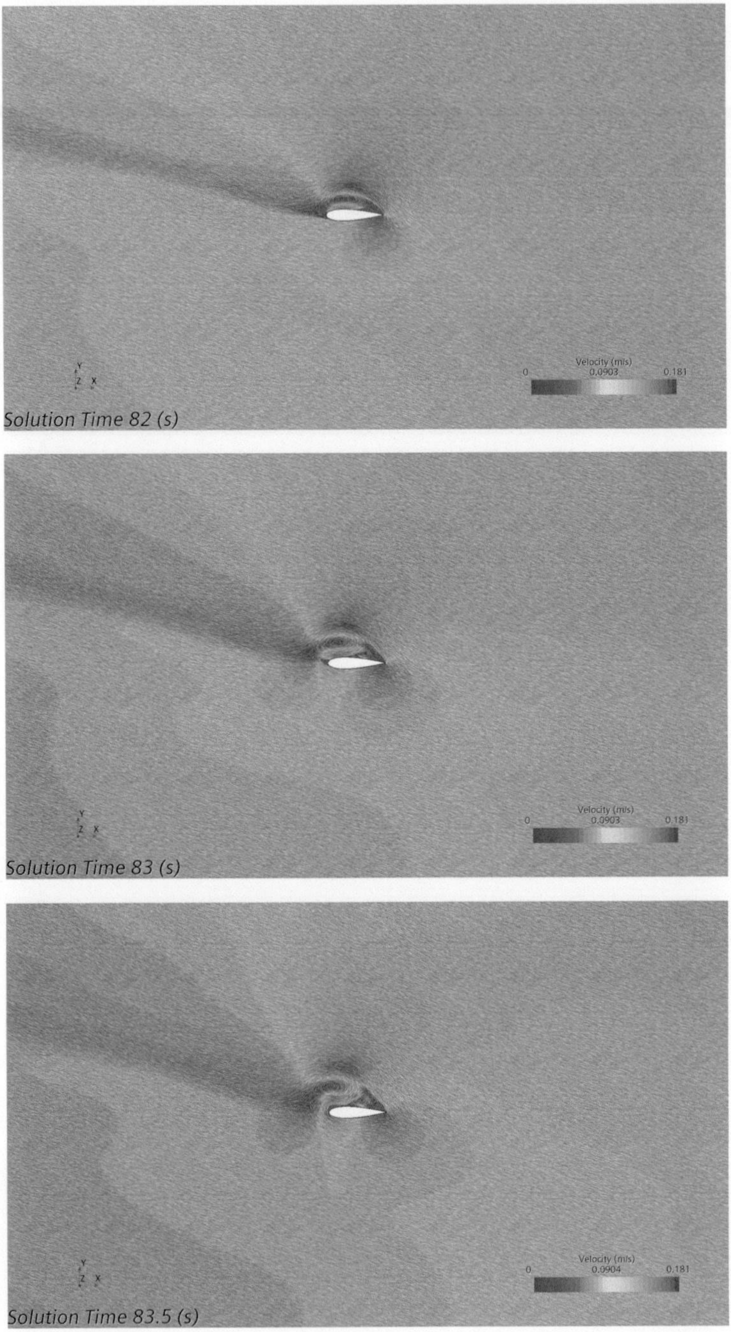

Fig. 7.16 (continued)

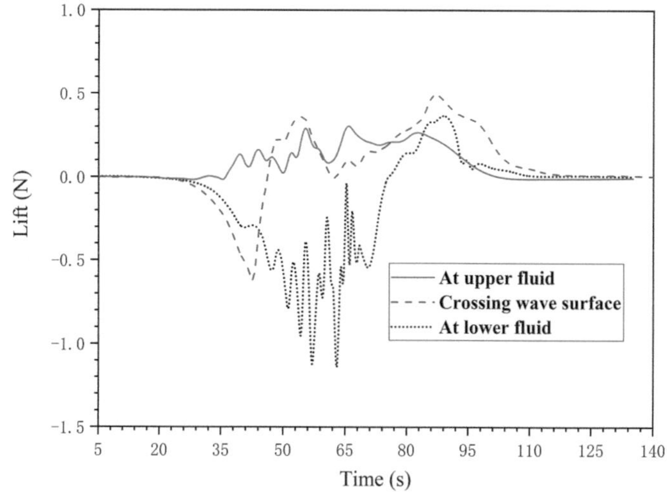

(a) Lift variation

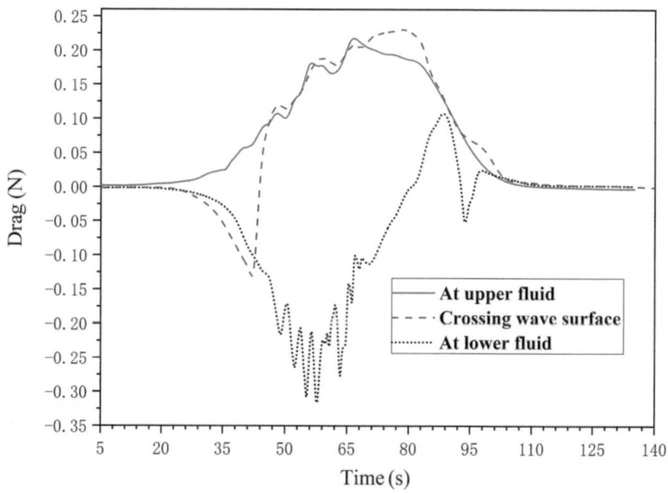

(b) Drag variation

Fig. 7.17 Interaction between internal solitary waves and 10° rudder structures at different submergence depths

At this point, the lift generated by the rudder angle reaches its peak. Subsequently, as the internal wave continues moving rightward, the horizontal velocity gradually decreases, and the lift diminishes accordingly. When the rudder structure approaches the left wave surface of the internal solitary wave, as mentioned earlier, the flow field in this region exhibits upward velocity. As the rudder structure crosses the internal solitary wave surface, the upward velocity gradually increases, leading to a corresponding rise in lift. For the rudder structure located below the internal solitary wave surface (Case 6), the presence of the rudder angle causes the oscillation amplitude of its lift variation curve to intensify during the phase below the right wave surface. By observing the velocity vector field during this phase, it is found that with the rudder angle, the range of flow separation induced by the rudder structure becomes larger, the flow field more complex, and vortex shedding more frequent, as shown in Fig. 7.18. This indicates that for a rudder located below the internal solitary wave surface, the presence of rudder angle during encounters with internal solitary waves significantly increases navigation risks compared to scenarios with 0° rudder angle.

Figure 7.17b shows the resistance variation curve of the rudder structure with a 10° rudder angle at different submerged depths under the action of internal solitary waves. It can be observed from the figure that the presence of the rudder angle increases the amplitude of the resistance on the rudder structure, which is consistent

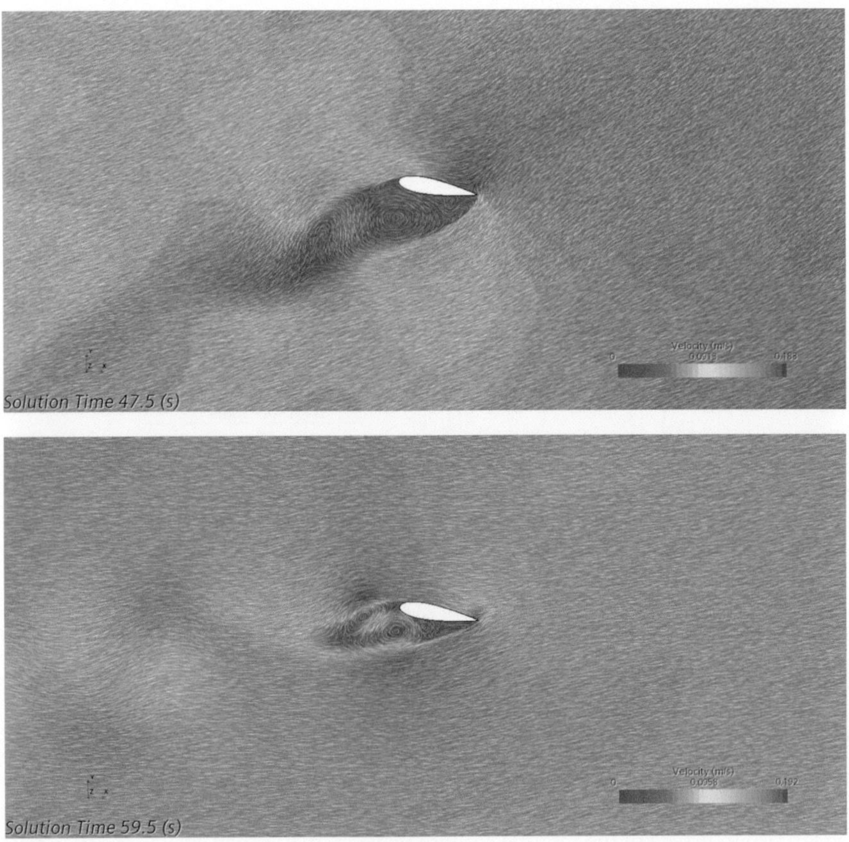

Fig. 7.18 Velocity vector diagram of the 10° rudder angle rudder structure below the right wave surface

with the phenomenon in uniform flow. The reasons for the changes in resistance values have been explained in detail in the previous analysis of the rudder structure with a 0° rudder angle, so they will not be repeated here. Notably, for the rudder structure located below the internal solitary wave surface (Case 6), when the left wave surface moves above the rudder structure, the oscillation degree of its resistance value decreases significantly. By analyzing the flow field during this stage, it is found that the periodic vortex structure previously described on the upper surface of the rudder no longer appears when the rudder angle exists, and the resistance amplitude also decreases significantly. The specific reasons will be discussed in detail in Sect. 7.3.3.

Through the study in this section, it can be found that when encountering internal solitary waves, the hydrodynamic performance of the rudder structure located above the wave surface (i.e., in the upper fluid) is minimally affected. Compared with the performance of the open-water rudder, only slight fluctuations occur near the corresponding values. For the rudder structure crossing the wave surface, the variation pattern of hydrodynamic performance is more obvious: when approaching the right wave surface, it is subjected to a downward vertical force, with the amplitude continuously increasing until reaching the maximum value upon crossing the right wave surface. Afterward, the trend of the rudder structure in the upper fluid is basically consistent—when approaching the left wave surface, it is subjected to an upward vertical force, with the amplitude also increasing until reaching the maximum value upon crossing the left wave surface. Special attention should be paid to the rudder structure located below the wave surface (i.e., in the lower fluid). The overall trend of hydrodynamic performance variation for the rudder structure in this position is similar to that of the rudder crossing the internal wave, except that it remains entirely in the lower fluid without entering the upper fluid phase. However, its force curve exhibits severe oscillations, and the force amplitude is much larger than that in other positions. Additionally, flow separation is highly prone to occur, posing a significant threat to the navigation safety of the vehicle.

7.3.3 Influence of Rudder Angle

The rudder angle is the most critical parameter affecting the hydrodynamic performance of the rudder structure itself. In the previous section, when the rudder angle changed, significant variations occurred in the lift, drag trends and amplitudes of the rudder structure. This section focuses on studying the influence of rudder angle on the interaction between internal solitary waves and rudder structures. In this section, the numerical wave tank for internal solitary waves is identical to that in Sect. 7.3.1, and the selected internal solitary wave model remains consistent with Sect. 7.3.2, with wave amplitude WA = 1.255 m, wave speed V = 0.3 m/s, and wavelength WL = 25.961 m. The submerged depth of rudder structures with different rudder angles remains consistent. According to the research results in Sect. 7.3.2, it can be observed that the variation patterns of hydrodynamic performance for rudder structures at different positions are significantly distinct under the influence of internal waves.

Therefore, to facilitate unified comparison, four rudder structures with different rudder angles are set at each of the following positions: above the wave surface, crossing the internal wave position, and below the wave surface.

(1) Interaction between internal solitary waves and rudder structures with different rudder angles located above the wave surface.

In the previous section's study, it was preliminarily found that for rudder structures located above the wave surface, the influence of rudder angle on their hydrodynamic performance variation patterns is limited. To further investigate the influence patterns and mechanisms of rudder angle, four rudder structures with different rudder angles were set at the same position above the wave surface, monitoring the changes in lift and drag under the action of internal solitary waves. The specific case settings are shown in Table 7.8.

Figure 7.19 shows the calculated results of lift and drag forces acting on rudder structures with different rudder angles located above the wave surface under the influence of internal solitary waves.

Figure 7.19a shows the lift force experienced by the rudder structure above the wave surface under the influence of internal solitary waves at rudder angles of 0°, 10°, 15°, and 20°, respectively. From the figure, it can be observed that the variation trends of lift force for different rudder angles are generally consistent—the larger the rudder angle, the greater the maximum lift force. The difference lies in the fact that when the rudder angle is 0° or 10°, the lift curve first decreases to a negative value before gradually rising. However, when the rudder angle exceeds 10°, the lift force no longer exhibits an initial decreasing trend but instead fluctuates and increases directly with the flow field velocity. As the internal solitary wave gradually moves away from the rudder structure, the lift force decreases to its initial value. The variation patterns and mechanisms for 0° and 10° rudder angles have been detailed in Sect. 7.3.2. When the rudder angle increases to 15°, the results show slight oscillations with relatively large amplitudes during the phase when the rudder structure is located above the right wave surface of the internal solitary wave, compared to the 0° and 10° cases. During this phase, the rudder structure is positioned around the middle of the right wave surface of the internal solitary wave. Analysis of the flow field information during this stage reveals that, compared to the 0° and 10° cases, the wake field of the rudder structure exhibits significant periodic vortex shedding (as shown in Fig. 7.20), leading to deformation of the wake field and a brief stall phenomenon

Table 7.8 Rudder angle case settings for rudder structures located above the wave surface

Case	Submerged depth d (m)	State	Rudder angle α	Wave amplitudeα (m)
Case 7	−0.5	Above wave surface	0°	1.225
Case 8	−0.5		10°	
Case 9	−0.5		15°	
Case 10	−3.0		20°	

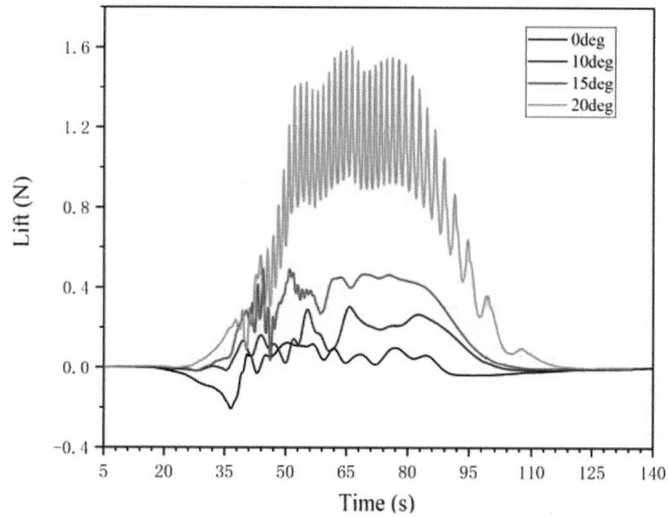

(a) Lift variation

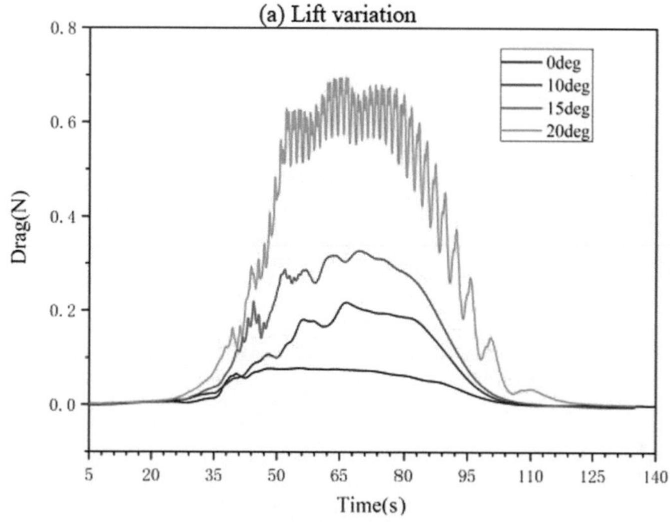

(b) Drag variation

Fig. 7.19 Interaction between internal solitary waves and rudder structures with different rudder angles located above the wave surface

(as shown in Fig. 7.21). After the rudder structure passes through this region, the wake field returns to normal, although some flow separation zones remain, but no further vortex shedding occurs. When the rudder angle further increases to 20°, the lift curve still shows an initial rise followed by a decline but with persistent and severe oscillations, indicating rudder failure. The stall angle advances to 20°, which could pose a fatal threat to the vessel in real-world conditions. Under this condition,

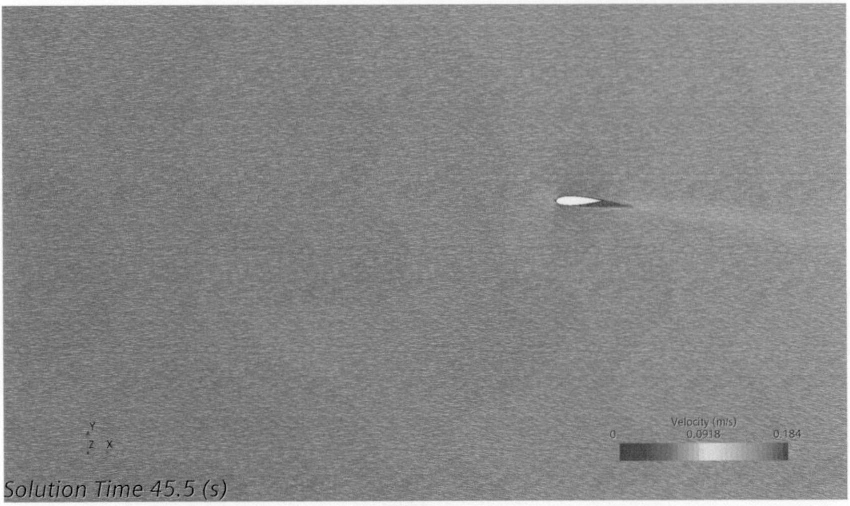

(a) 0° rudder angle

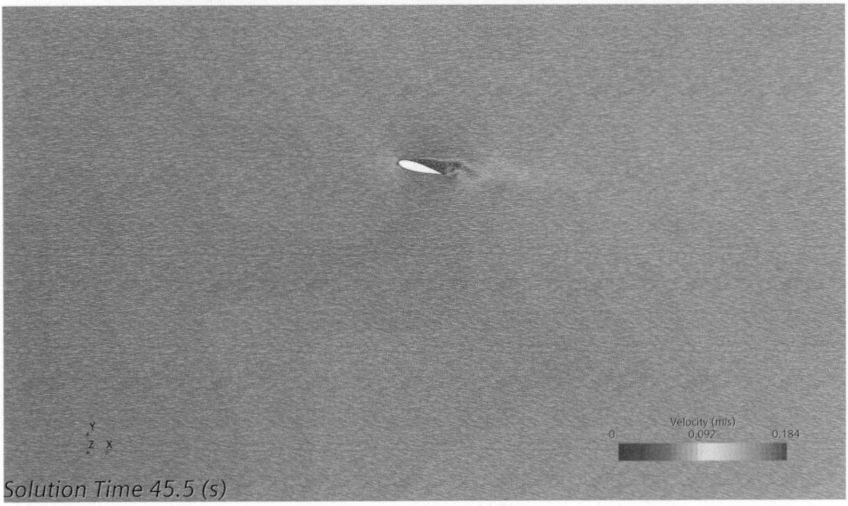

(b) 15° rudder angle

Fig. 7.20 Velocity vectors of rudder structures with different angles above the wave surface

vortex shedding in the wake field of the rudder structure begins continuously as soon as the internal solitary wave approaches and persists until the wave moves away, as shown in Fig. 7.22.

Figure 7.19b presents the drag force experienced by the rudder structure above the wave surface under the influence of internal solitary waves at rudder angles of 0°, 10°, 15°, and 20°, respectively. From the figure, it can be seen that the variation trends of drag force for different rudder angles are almost identical, differing only in

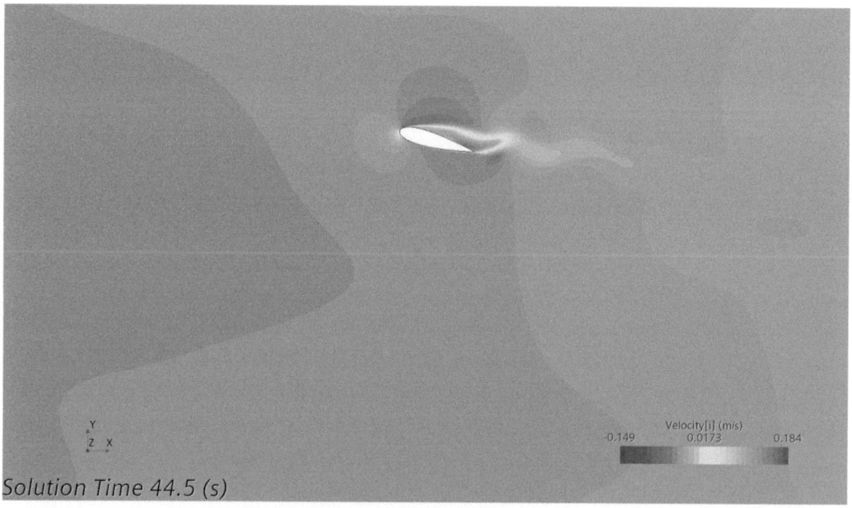

(a) 44.5s Horizontal velocity field

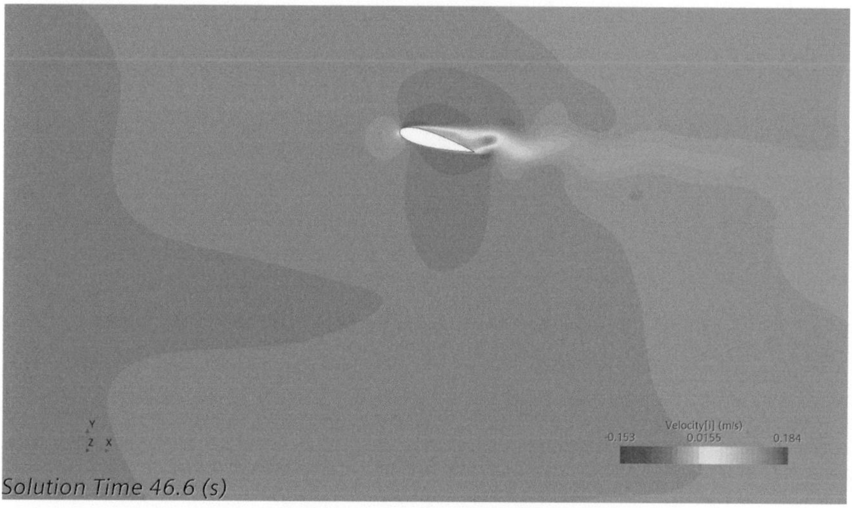

(b) 46.6s Horizontal velocity field

Fig. 7.21 Horizontal velocity field of rudder structure with 15° angle located above wave surface

oscillation intensity and amplitude. The larger the rudder angle, the more pronounced the oscillations and the greater the extreme values of drag force. Therefore, the trends and reasons for drag variation can be referred to the descriptions and analyses of Case 1 and Case 4 in Sect. 7.2.2. In the case of a 15° rudder angle, the oscillations in drag force between 42 and 50 s are attributed to the same cause as the lift oscillations—a brief stall of the rudder structure.

Fig. 7.22 Velocity vectors of rudder structure with 20° angle located above wave surface

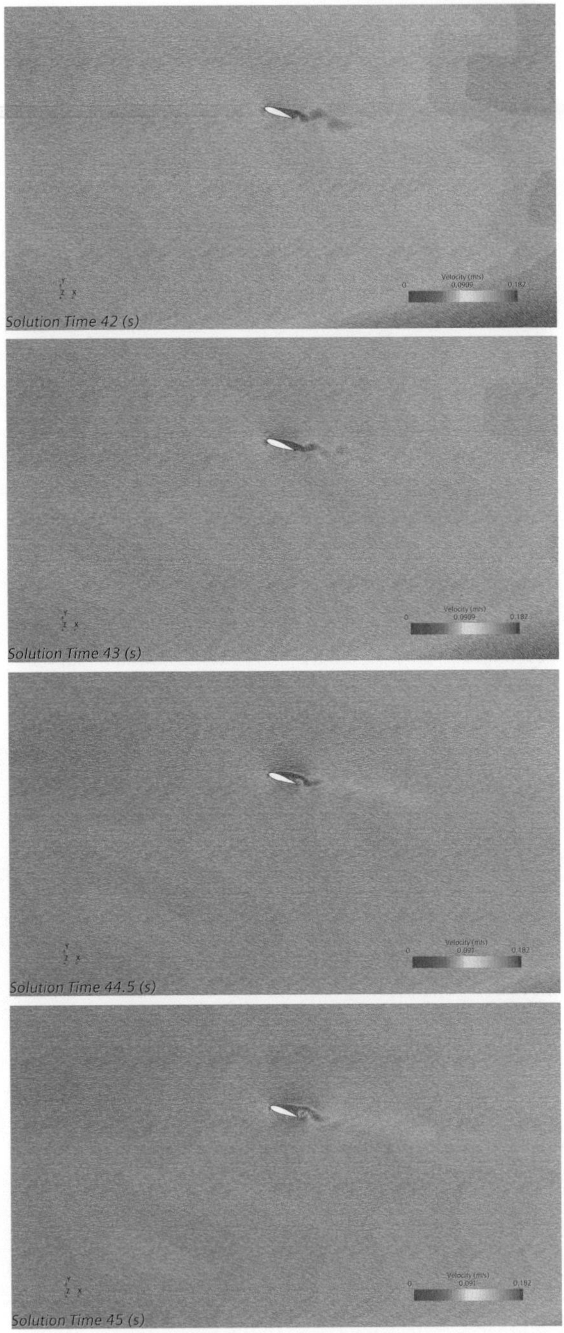

(2) Interaction between internal solitary waves and rudder structures with different angles crossing wave surface.

Based on previous research, four rudder structures with different angles were set at the same position crossing the wave surface to monitor the changes in lift and drag forces under internal solitary waves. Specific working condition settings are shown in Table 7.9.

Figure 7.23 shows the calculated results of lift and drag forces on rudder structures with different angles crossing wave surface under internal solitary waves.

Figure 7.23a presents the lift forces acting on rudder structures with different angles (0°, 10°, 15°, and 20°) when crossing the wave surface under internal solitary waves. The graph reveals that the variation trends of lift forces for rudders with different angles are highly consistent, differing only in the magnitude of forces. The larger the rudder angle, the greater the maximum upward lift force. All rudder angles show a similar trend of initial decrease in lift force due to the downward vertical force generated by the clockwise-rotating velocity field in the flow. As the rudder angle increases, the negative vertical force decreases because when the rudder is near the right wave surface of the internal solitary wave, the fluid flows toward the lower left. A larger rudder angle reduces the projected area of the flow-facing surface on the horizontal plane, thereby decreasing the negative vertical force. When the right wave surface begins to cross the rudder structure, the lift force increases with the horizontal velocity. Upon entering the upper fluid layer, where the velocity is mainly in the positive X-direction, larger rudder angles result in greater maximum lift forces. After the wave trough passes the rudder, the horizontal velocity decreases, and so does the lift force. When the left wave surface starts crossing the rudder, the lift force rises again. Except for the 20° rudder angle case, the other three cases reach nearly identical maximum values during this second rise. Finally, as the internal solitary wave moves away from the rudder, the lift force gradually returns to its initial value. Notably, when the rudder is at the wave-crossing position, the 15° rudder angle case does not experience stalling; instead, the lift curve becomes smoother, indicating that rudders above the wave surface are more sensitive to angle changes compared to those crossing the wave surface. Consistent with the above-wave-surface cases, when the rudder angle increases to 20°, the rudder structure experiences stalling, with flow separation occurring in the flow field. The wake field of the rudder begins to continuously shed vortex structures, as shown in Fig. 7.24. However, unlike the above-wave-surface cases, stalling occurs later for the wave-crossing rudder, around

Table 7.9 Working condition settings for rudder structures crossing wave surface

Case	Submergence depth d (m)	State	Rudder angle α	Wave amplitude a (m)
Case 11	−1.5	Crossing wave surface	0°	1.225
Case 12			10°	
Case 13			15°	
Case 14			20°	

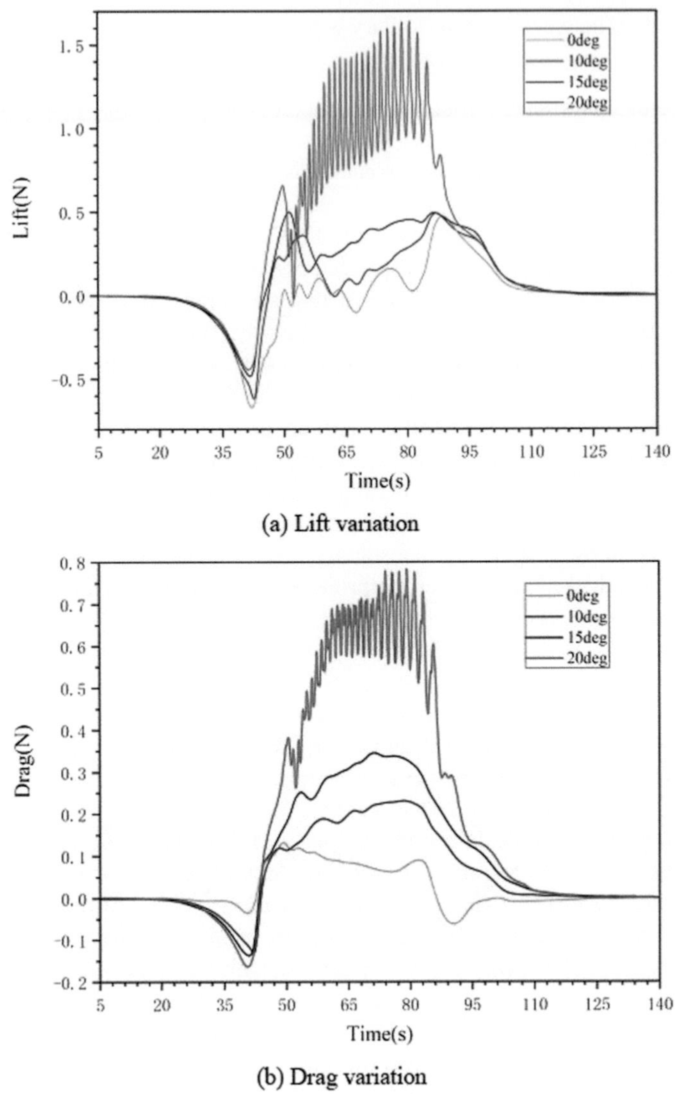

(a) Lift variation

(b) Drag variation

Fig. 7.23 Interaction between internal solitary waves and rudder structures with different angles located at wave surface crossing position

50 s, i.e., only after the rudder has entered the upper fluid layer does flow separation occur.

Figure 7.23b shows the resistance variation curves of rudder structures with angles of 0°, 10°, 15°, and 20° crossing the wave surface under internal solitary wave action. It can be observed that the resistance variation trends for different rudder angles are generally consistent, differing mainly in magnitude—the larger the rudder angle,

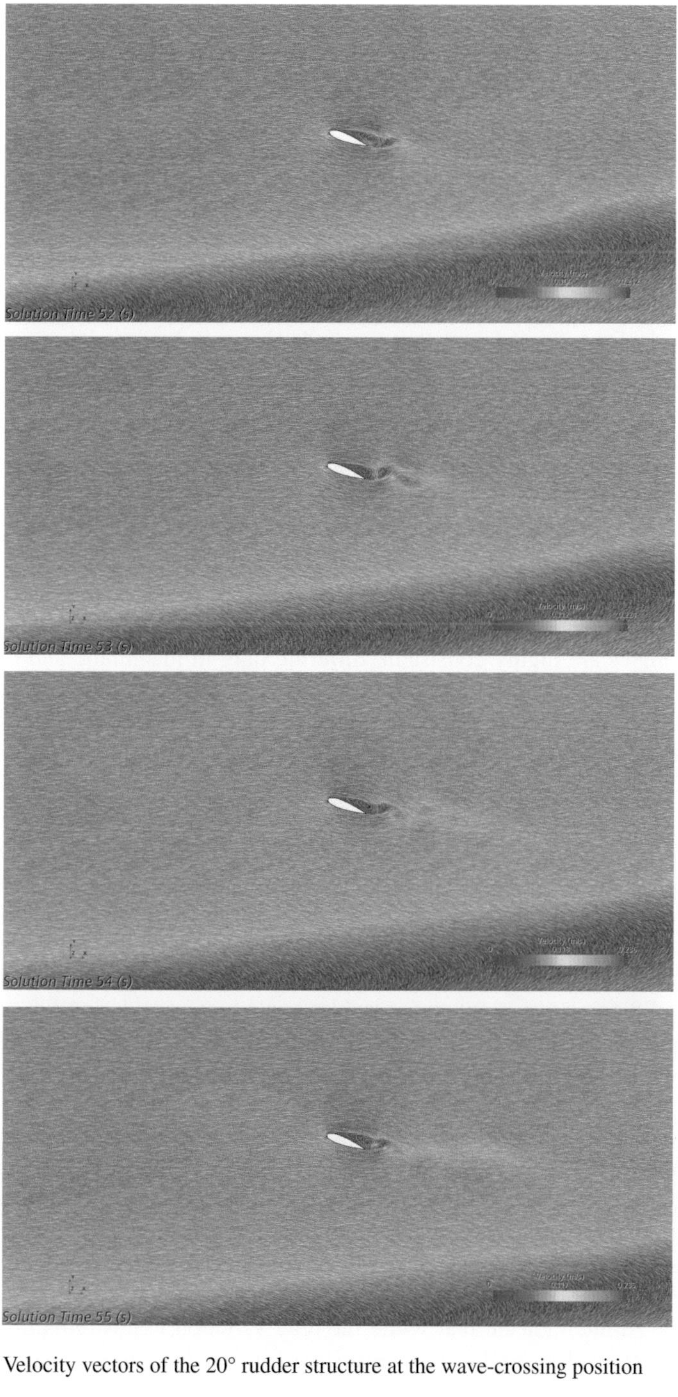

Fig. 7.24 Velocity vectors of the 20° rudder structure at the wave-crossing position

the greater the resistance. The trends and reasons for resistance variation can be referenced to the description and analysis of Case 2 and Case 5 in Sect. 7.2.2. The only difference occurs during the 80s–95s phase when the rudder structure approaches and crosses the left wave surface of the internal solitary wave: the 0° rudder angle structure shows a directional change in resistance, which doesn't occur in the other three cases. This is due to the vortex structure appearing on the upper surface of the 0° rudder, which disappears when there's a rudder angle.

(3) Interaction between rudder structures with different angles located below the wave surface and internal solitary waves.

The study in Sect. 7.3.2 found that the hydrodynamic load variations on rudder structures below the wave surface are relatively complex. Four rudder structures with different angles were placed at the same position below the wave surface to monitor lift and resistance changes under internal solitary wave action. Specific case settings are shown in Table 7.10. Figure 7.25 presents the calculated results of lift and resistance forces on rudder structures with different angles crossing the wave surface under internal solitary wave action.

Figure 7.25a presents the lift forces acting on the rudder structure with different angles (0°, 10°, 15°, and 20°) crossing the wave surface under internal solitary wave conditions. The results show that the variation trends of lift forces for rudders with different angles are highly consistent, differing only in magnitude and oscillation intensity. All cases exhibit an initial oscillatory decrease with downward vertical force, followed by a gradual increase. The changing trends and underlying mechanisms can be referenced to the descriptions and analyses of Case 3 and Case 6 in Sect. 7.2.2. Comparing the four curves reveals that larger rudder angles correspond to greater maximum downward vertical forces and smaller maximum upward lift forces. Moreover, larger rudder angles lead to more severe oscillations in the force curves. Combined with flow field nephograms, it's observed that increased rudder angles prolong flow separation duration and expand its influence on the flow field, as demonstrated in Fig. 7.26's comparative velocity vector diagrams of different rudder angles at the same time. The distinctive feature among the four working conditions lies in the lift force variation: as the rudder angle increases during the process of the rudder approaching and moving away from the left wave surface, except for the 0° case, the lift force no longer shows significant upward surges. This phenomenon occurs because when the rudder is located beneath the left wave surface of the internal solitary wave, the surrounding flow moves toward the upper left, generating upward lift. However, with increasing rudder angle, the angle between flow direction and the rudder's cross-section decreases, reducing the horizontal projection area of the flow-facing surface and consequently diminishing the vertical force.

Figure 7.25b shows the drag variation curves of rudder structures with different angles (0°, 10°, 15°, and 20°) crossing the wave surface under internal solitary wave action. The results indicate that while the drag variation trends remain consistent across different rudder angles, differences exist in force magnitude and oscillation intensity. All cases initially experience a decrease in drag (negative X-direction force), with larger rudder angles corresponding to greater drag forces-consistent

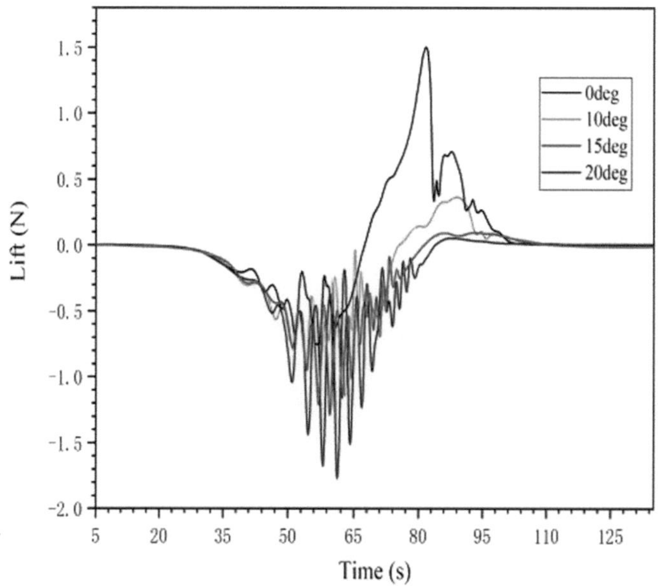

(a) Lift variation

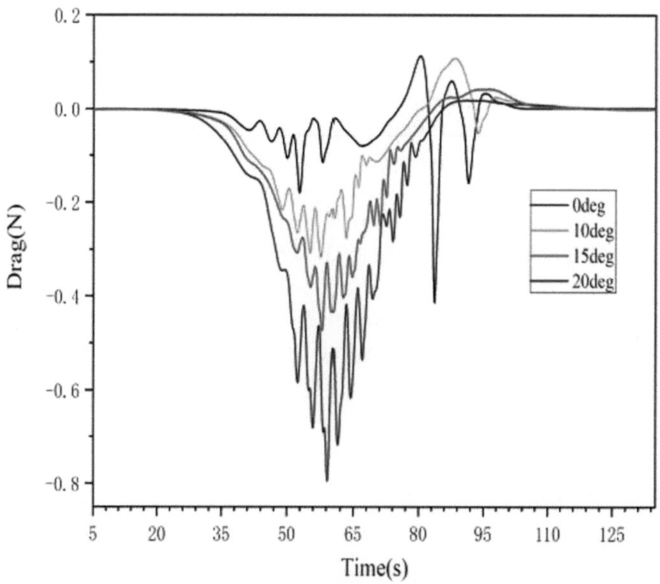

(b) Resistance variation

Fig. 7.25 Interaction between internal solitary waves and rudder structures with different angles located below the wave surface

Table 7.10 Rudder angle case settings for structures below the wave surface

Case	Submergence depth d (m)	State	Rudder angle α	Wave amplitude a (m)
Case 15	−3.0	Below wave surface	0°	1.225
Case 16			10°	
Case 17			15°	
Case 18			20°	

with open-water rudder behavior in uniform flow. Subsequently, the drag gradually increases, with the trend and underlying mechanisms explained in Sect. 7.2.2 (Case 3 and Case 6 analysis). The primary distinction appears in the 0° case around 83s, where an additional trough occurs with abrupt directional change in drag force and relatively large absolute value, unlike the other three cases. As explained in Sect. 7.2.2, this phenomenon stems from periodic vortex structures forming on the rudder's upper surface, causing localized flow reversal (positive X-direction) in surrounding areas. Vortex shedding induces rapid flow direction changes, leading to sudden drag reversals. With increasing rudder angles, the streamline angle relative to the cross-section decreases significantly, reducing flow obstruction and eliminating periodic vortex formation on the upper surface.

This study reveals that in internal wave environments, rudder hydrodynamic performance shows significant sensitivity to angle variations. Regardless of position relative to the solitary wave, stall angles occur substantially earlier, necessitating extreme caution during rudder operations. Taking this solitary wave model as example: (1) For rudders above wave surface, operational angles should remain below 15°, preferably around 10°—as 15° already exhibits localized stall; (2) For rudders crossing wave surface, maximum operational angle should be controlled near 15°, with stall onset delayed compared to above/below wave positions; (3) Most critically, rudders beneath wave surface experience localized stall even at 0° angle, with persistent stall occurring at any positive angle. Increasing angles prolong flow separation duration and expand affected flow regions. Therefore, attitude adjustments via rudder should be strictly avoided when positioned below wave surface.

7.3.4 Wave Amplitude Effects

As one of the important parameters of internal solitary waves, wave amplitude also has certain influence on the hydrodynamic performance of rudder structures. Therefore, this section focuses on studying the impact of wave amplitude on the interaction between internal solitary waves and rudder structures. In this section, the numerical wave tank for internal solitary waves is completely consistent with that in Sect. 7.3.1, and the internal solitary wave model settings are also identical, with only the wave amplitude being changed. Three different amplitudes of internal solitary waves were

Fig. 7.26 Velocity vectors of different-angle rudder structures beneath wave surface at identical time instants

(a) 0° rudder angle

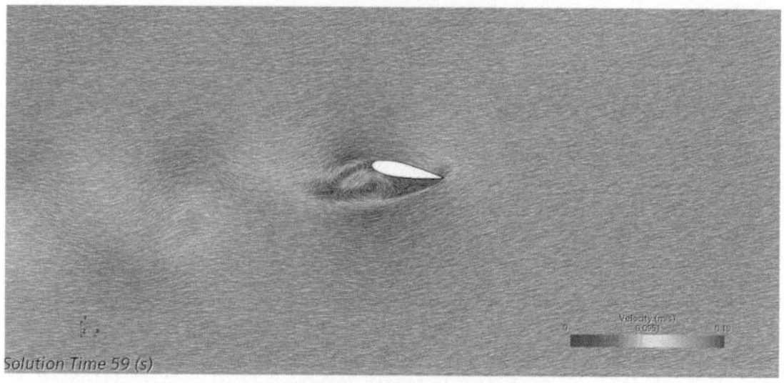

(b) 10° rudder angle

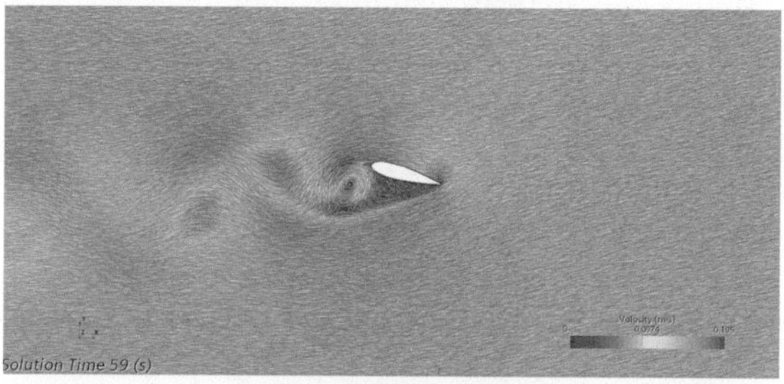

(c) 15° rudder angle

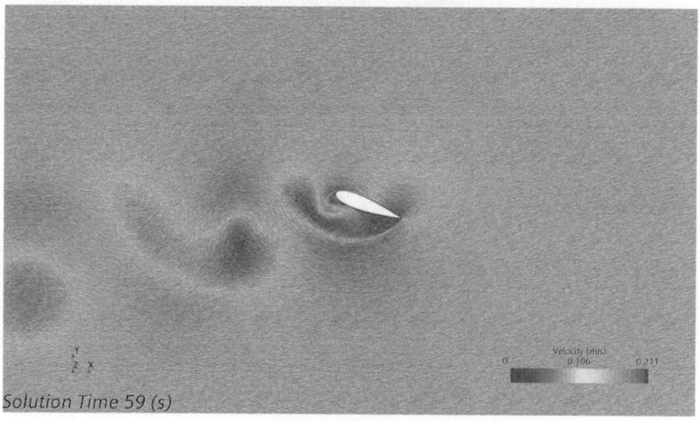

(d) 20° rudder angle

selected to study the influence pattern of wave amplitude on hydrodynamic characteristics when rudder structures traverse the wave surface. Previous research has shown that the trends of results may differ between cases with 0° rudder angle and those with existing rudder angles. Therefore, this section simulates scenarios of rudder structures with 0° and 10° angles traversing internal solitary wave surfaces under three different wave amplitude conditions. The specific case settings are shown in Table 7.11.

(1) Influence of wave amplitude on hydrodynamic performance when 0° rudder angle structure traverses wave surface

This section focuses on working conditions Cases 19–21, simulating three internal solitary waves with different amplitudes at the same submergence depth for the rudder structure, while monitoring the changes in lift and drag forces under the action of internal solitary waves. Figure 7.27 presents the calculated results of lift and drag forces acting on the rudder structure crossing the wave surface under internal solitary waves with different amplitudes.

Figure 7.27a shows the lift force acting on the 0° rudder-angle rudder structure crossing the wave surface when interacting with internal solitary waves of three different amplitudes (0.990 m, 1.102 m, and 1.255 m). It can be observed that the trends of lift force variation under different wave amplitudes are almost identical, differing only in phase and force amplitude. The larger the amplitude of the internal solitary wave, the greater the wavelength and wave velocity, causing the force curve to shift earlier. Additionally, the peak values of the vertical force increase, and the fluctuation amplitude of the curve also becomes larger. The same applies to the drag curves obtained in Fig. 7.27b, where the trends under the three amplitude conditions are completely consistent. The variation trends and underlying reasons for the lift and drag forces acting on the rudder structure crossing the wave surface have been explained in detail earlier, and reference can be made to the discussion in Case 2.

(2) Influence of wave amplitude on hydrodynamic performance of 10° rudder angle structure crossing wave surface

This section focuses on working conditions Cases 22–24, simulating three internal solitary waves with different amplitudes acting on a rudder structure with 10° rudder

Table 7.11 Case settings for wave amplitude influence study

Case	Submergence depth d (m)	State	Rudder angle a	Wave amplitude a (m)
Case 19	−1.5	Traversing wave	0°	0.9904
Case 20				1.1020
Case 21				1.2550
Case 22			10°	0.9904
Case 23				1.1020
Case 24				1.2550

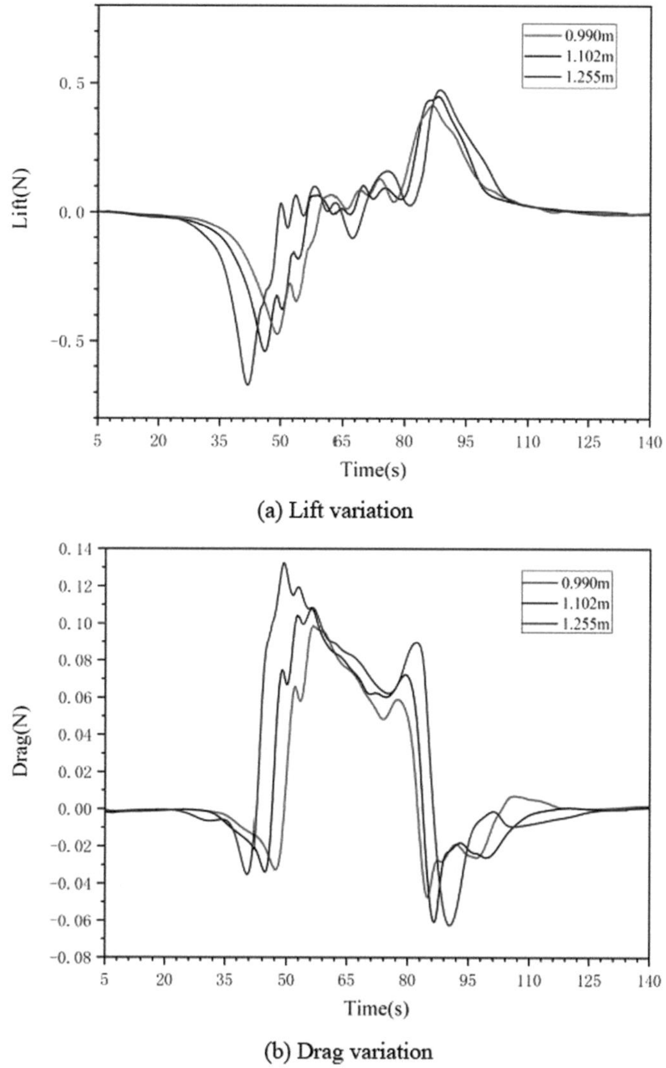

Fig. 7.27 Interaction between internal solitary waves of different amplitudes and the rudder structure crossing the wave surface

angle at the same submerged depth, and monitoring the variations in lift and drag forces. Figure 7.28 presents the calculated results of lift and drag forces on the 10° rudder structure crossing the wave surface under different internal solitary wave amplitudes.

From Fig. 7.28, it can be observed that under different wave amplitudes, the force variation trends on the 10° rudder structure are completely consistent, with differences only in phase and force amplitude. Larger wave amplitudes result in

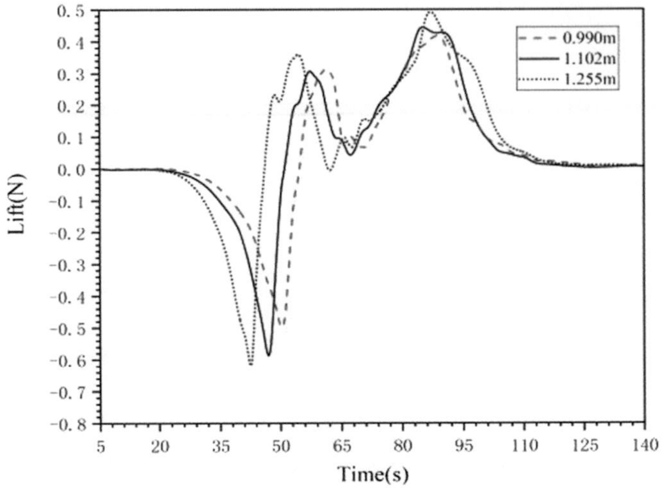

(a) Lift variation

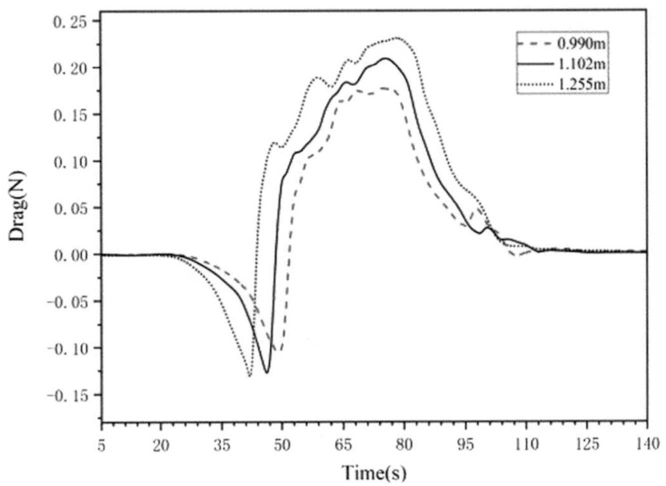

(b) Drag variation

Fig. 7.28 Interaction between internal solitary waves with different amplitudes and a 10° rudder structure crossing the wave surface

greater maximum forces, but compared to the 0° rudder angle case, the hydrodynamic performance of the 10° rudder structure is less sensitive to changes in internal wave amplitude, with minimal variation in peak forces. The trends and reasons for lift and drag variations on the 10° rudder structure crossing the wave surface have been explained in detail earlier and can be referenced in Case 5.

The above research reveals that wave amplitude has limited influence on the variation patterns of internal solitary wave forces acting on the rudder structure, with changes mainly occurring in phase, peak forces, and curve fluctuation amplitudes.

7.4 Summary

Based on the continuously stratified internal solitary wave theory—the DJL equation—this chapter establishes a numerical wave tank using the interpolation wave-making method to simulate interactions between internal solitary waves and rudder structures. After verifying grid independence and the accuracy of the numerical simulation method, the study investigates the influence patterns of internal solitary waves on the hydrodynamic performance of rudder structures, focusing primarily on submerged depth, rudder angle, and wave amplitude. The research indicates:

(1) Hydrodynamic loads on rudder structures at different positions are affected differently by internal solitary waves. The hydrodynamic performance of rudder structures above the wave surface is minimally affected; for rudder structures crossing the wave surface, the variation patterns in hydrodynamic performance are more pronounced, with downward vertical forces increasing in amplitude when approaching the right wave surface until reaching maximum values upon crossing, followed by similar trends in the upper fluid layer, where upward vertical forces increase in amplitude when approaching the left wave surface until reaching maximum values upon crossing. Special attention is required for rudder structures below the wave surface, as their force curves exhibit severe oscillations with significantly larger amplitudes than other positions, and flow separation is highly likely to occur, posing substantial threats to the navigation safety of underwater vehicles.

(2) The influence of rudder angle is most significant. Regardless of the rudder structure's position relative to the internal solitary wave, the stall angle of the rudder structure will be significantly advanced, indicating that rudder control in an internal wave environment must be extremely cautious. If the rudder structure is located above the wave surface, the steering angle should be kept below 15°, preferably around 10°. If the rudder structure is positioned traversing the wave surface, the maximum steering angle should be controlled around 15°. Special attention must be paid when the rudder structure is located below the wave surface—even at a rudder angle of 0°, a small-scale stall phenomenon occurs. When a rudder angle is applied, continuous stalling occurs. Therefore, when the rudder structure is below the wave surface, attitude adjustments via steering should be avoided as much as possible.

(3) The wave amplitude has little effect on the variation pattern of the internal solitary wave forces acting on the rudder structure. The main differences lie in the phase, peak forces, and amplitude of the force curve fluctuations.

Chapter 8
Self-propulsion Experiments of Submersible Under the Action of Internal Waves

8.1 Introduction

In the previous two chapters, numerical simulations were used to study the interaction between internal solitary waves and two key appendages of underwater vehicles—propellers and rudders—revealing significant influences and certain operational patterns. To further verify whether the conclusions drawn from the research can provide practical maneuvering strategies in real environments, additional studies are necessary. Physical experiments are one of the important means for investigating the interaction between internal waves and underwater vehicles. However, due to limitations in laboratory wave tank dimensions, vehicle size, and post-processing, most current research on the interaction between internal solitary waves and underwater vehicles relies on numerical simulations. The few experimental studies on internal solitary waves and vehicle interactions have used fixed or towed vehicles, allowing only analysis of flow fields or vehicle forces, making it difficult to obtain the true motion response of vehicles under internal wave conditions. Moreover, these studies cannot explore the influence of different appendage control strategies on vehicles in an internal solitary wave environment.

Based on the understanding of internal wave effects on underwater vehicles and their appendages, this chapter designs physical experiments to study the interaction between internal waves and both suspended and self-propelled vehicles. A scaled vehicle model matching the internal solitary wave experimental flume and the generated wave parameters is constructed. Self-propulsion experiments of underwater vehicles in an internal solitary wave environment are conducted to partially validate the aforementioned conclusions, providing valuable references for the safe underwater navigation of vehicles in real oceanic conditions.

P. Du et al., *Internal Wave Hydrodynamics on Submersibles*,
https://doi.org/10.1007/978-981-95-4479-0_8

8.2 Self-propulsion Experiment Design

To obtain accurate experimental results, a comprehensive experimental design must first be conducted. This section presents the physical experimental design for the interaction between internal solitary waves and both suspended and self-propelled underwater vehicles. Based on the dimensions of the internal solitary wave tank and the parameters of the generated internal solitary waves, a scale-matched underwater vehicle model was designed and fabricated. Internal solitary waves were generated using the gravity collapse method, and the control variable method was employed to design working conditions by adjusting parameters such as wave amplitude, vehicle state (suspended/self-propelled), vehicle position, speed, and rudder angle. The design concept is illustrated in Fig. 8.1.

8.2.1 Internal Solitary Wave Experimental Tank

(1) Stratified Flow Experimental Tank

The prerequisite for generating internal solitary waves is the presence of a density-stratified fluid environment. The experimental research in this chapter was conducted in a gravity-type stratified flow experimental tank, which consists of five main components: the evolution tank body, stratified fluid preparation device, internal solitary wave generation device, wave absorption device, and experimental recording device, as shown in Figs. 8.2 and 8.3. The tank is divided into a wave generation zone, working zone, and wave absorption zone. The tank is made of tempered glass, with a total length of 4500 mm, width of 500 mm, and height of 600 mm, allowing for a maximum water depth of 500 mm. There are two mushroom-shaped water inlets at the bottom of the tank. The width of the wave generation zone can be adjusted and is separated from the working zone by a gate.

(2) "Dual-Tank Method" for Preparing Stratified Fluid

First, it is necessary to prepare a stable stratified fluid environment. This study uses a stratified fluid preparation device based on the "dual-tank method," which mainly consists of a saltwater tank, freshwater tank, stirring device, and peristaltic pump, as shown in Fig. 8.4. First, the required volumes of freshwater and saltwater are calculated based on the tank parameters and internal solitary wave parameters. Saltwater of the corresponding density is prepared in the saltwater tank according to the desired stratified environment parameters. The tank is equipped with a dual-blade stirring device to accelerate the mixing of saltwater, with a blade diameter of 20 cm, and its rotation speed can be adjusted via a control device. After the saltwater is prepared, the fluid is slowly injected into the tank through the mushroom-shaped inlets using the peristaltic pump—first freshwater, then saltwater. The mushroom-shaped inlets allow the lower-layer fluid to slowly lift the upper-layer fluid, thereby

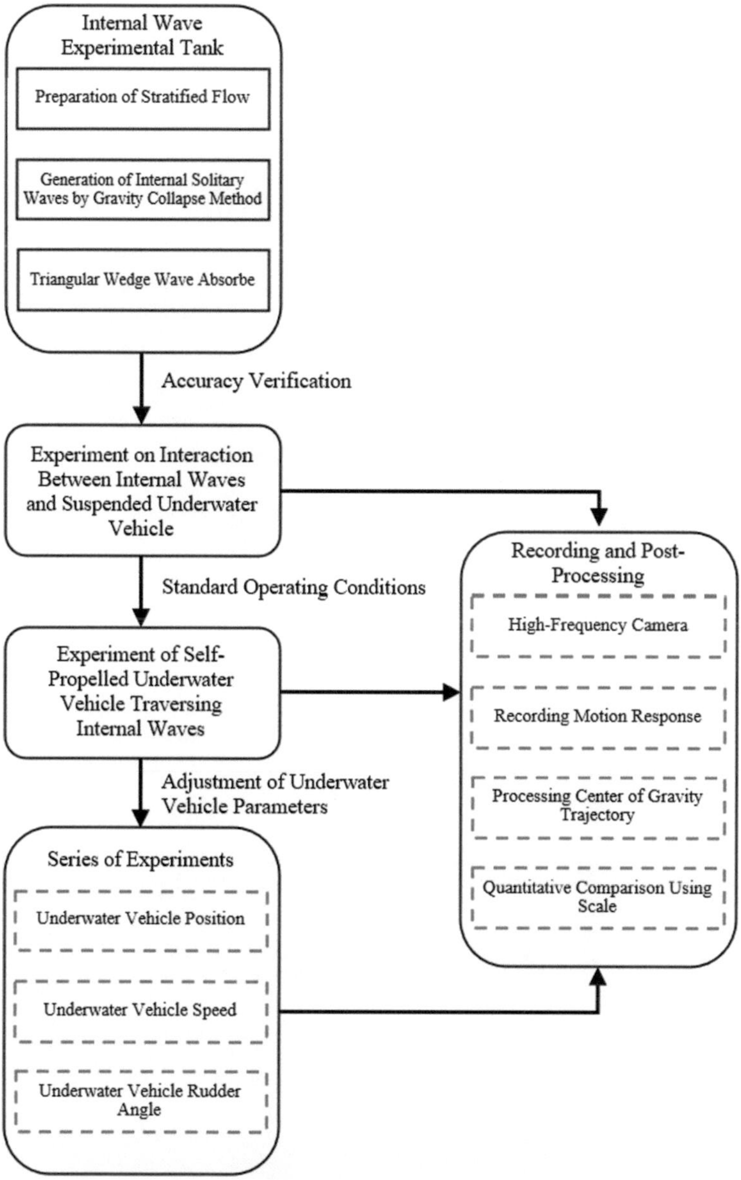

Fig. 8.1 Design concept for self-propelled experiments in an internal solitary wave environment

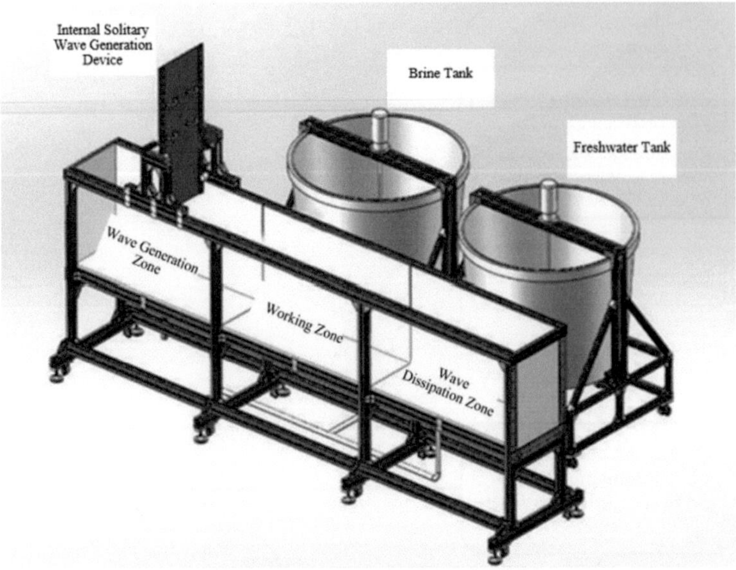

Fig. 8.2 Schematic diagram of the stratified flow experimental tank

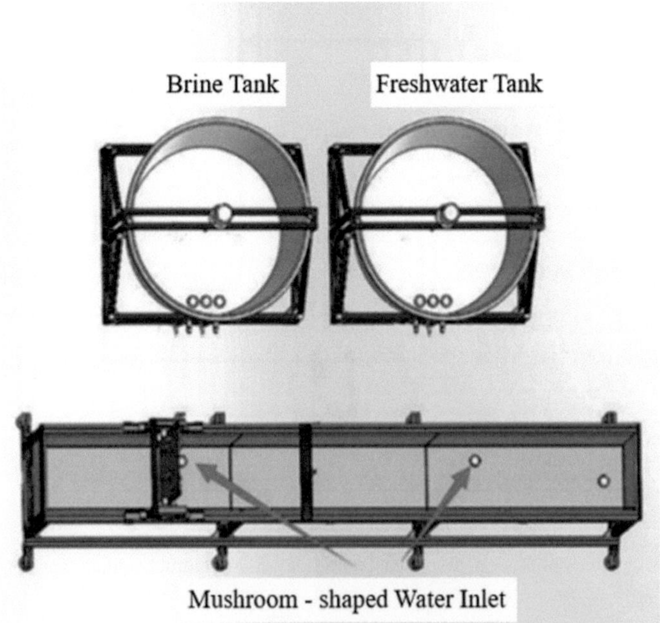

Fig. 8.3 Schematic diagram of the stratified flow preparation device

reducing mixing between the saltwater and freshwater. The injection speed of the fluid can be set via the peristaltic pump's control system.

For a strongly stratified two-layer fluid system, the fluids can be injected sequentially, resulting in the stratified environment shown in Fig. 8.4a. For a three-layer fluid system, fresh water is injected first, followed by injecting saline water into the fresh water tank. After thorough mixing, the mixture is injected into the flume, and finally, saline water is injected again. For a continuous density fluid system, fresh water is injected first, then saline water is injected into the fresh water tank while simultaneously injecting the mixed fluid from the fresh water tank into the flume. During this process, by adjusting the injection rate of saline water into the fresh water tank, a density distribution satisfying the hyperbolic tangent function can be achieved, followed by injecting the remaining saline water.

(a) Two-layer stratified fluid environment

(b) Continuous stratified fluid environment

Fig. 8.4 Stratified fluid environment

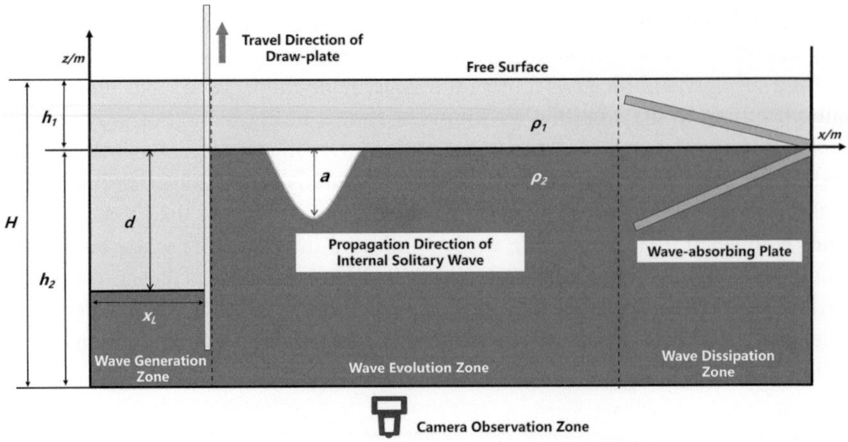

Fig. 8.5 Schematic diagram of the internal solitary wave experiment

In the experimental research of this chapter, all prepared stratified fluid environments were strongly stratified two-layer systems. The upper layer fluid density was set to (fresh water), and the lower layer fluid density was set to (saline water). The total fluid height was 50 cm, with the upper layer (fresh water) height being and the lower layer (saline water) height being. The prepared stratified flow environment is shown in Fig. 8.5.

(3) Internal Solitary Wave Generation and Absorption Devices

After establishing a stable stratified flow environment, the internal solitary wave generation device can be used to produce the required internal solitary waves. Both ends of the stratified flow experimental flume are equipped with internal solitary wave generation and absorption devices, as shown in Fig. 8.5.

This experimental study employed the gravitational collapse method to generate internal solitary waves, using a draw-plate method as the wave generation device (Fig. 8.6a). The specific implementation is as follows: After completing the preparation of the density-stratified fluid system, a draw-plate is used to separate the stratified fluid in the flume into left and right sections, without touching the bottom (maintaining connection between the two regions at the bottom layer). This creates a communicating vessel system. The upper layer fluid (fresh water) in the working area is pumped into the wave generation area. Since the two sides of the draw-plate are connected and pressure must remain balanced, the injected fresh water in the wave generation area will push the underlying saline water into the working area, creating a height difference between the density interfaces of the two-layer fluids on either side of the draw-plate. During this process, a plastic film can be placed on the surface of the wave generation area to minimize fluid disturbance. Due to the lower density of fresh water, there exists a small free surface height difference between the two sides of the draw-plate, while the height difference at the density interface is much larger—this being the key factor affecting the amplitude of internal solitary waves

(Fig. 8.7a). In this experiment, the height difference of the density interface between the two-layer fluids on either side of the draw-plate serves as the control variable. When the present height difference is reached, fresh water pumping is stopped, and the draw-plate is quickly withdrawn upward manually. Under gravity, this action forms a stably propagating internal solitary wave (Fig. 8.7b).

The wave-absorbing device in this experiment consists of a triangular wedge structure composed of a group of V-shaped plates densely distributed with small holes. The V-shaped plate is 150 cm long and 45 cm wide, as shown in Fig. 8.6b.

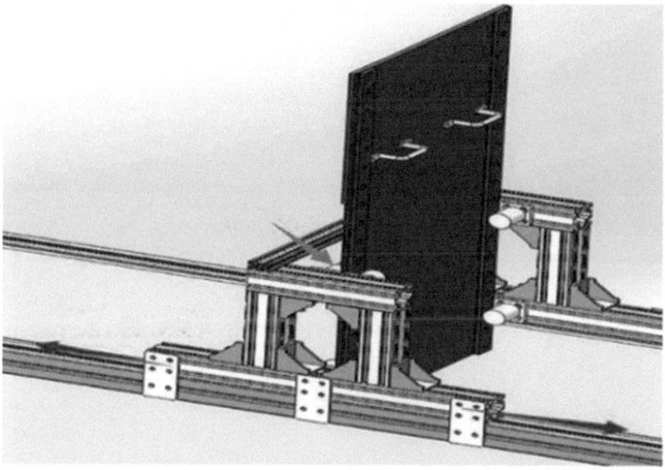

(a) "Drawplate method" wave-making device

(b) Triangular wave-absorbing device

Fig. 8.6 Schematic diagram of the internal solitary wave experiment

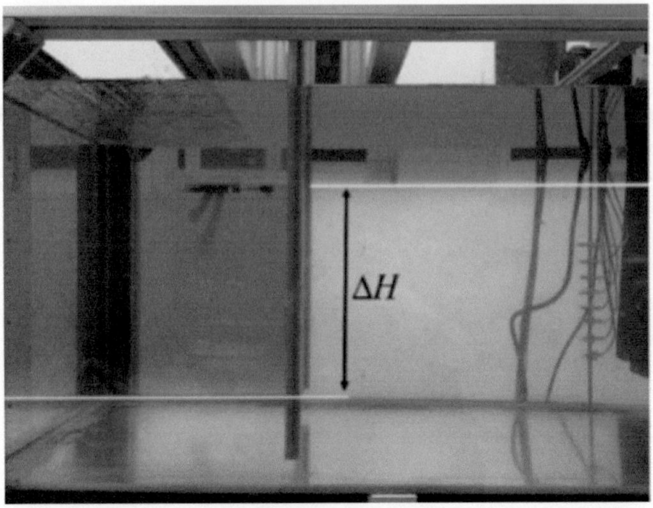

(a) Gravity collapse area

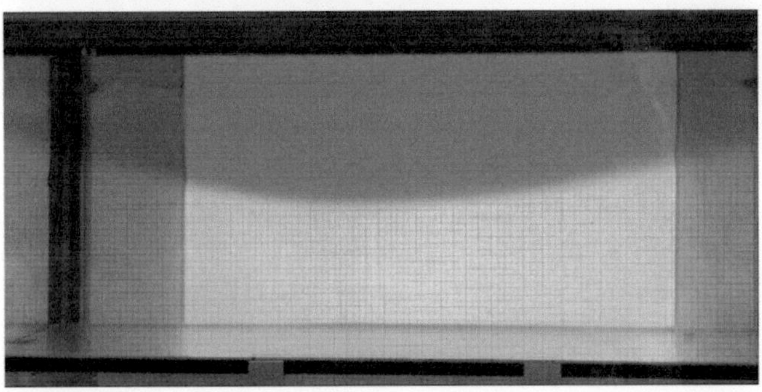

(b) Strongly stratified internal solitary wave generated in the experiment

Fig. 8.7 Laboratory wave-making for internal solitary waves

The small holes in the V-shaped plate allow fluid to pass through smoothly without affecting the construction of stratified flow. The two ends of the two wave-absorbing plates are connected to the connecting frames at the tail and side of the water tank through fixed clamps, respectively. The apex angle position and the angle of the V-shaped plate can be adjusted according to the position of the density interface and the size of the wave amplitude.

The picture contains a game console, a cart, and a table Automatically generated description The picture contains a game console, a cart, and a table Automatically generated description.

8.2.2 Design and Fabrication of Self-navigating Vehicles

After completing the laboratory wave-making for internal solitary waves, it is necessary to determine the functions that the navigation body model needs to achieve based on the size of the experimental water tank, the scale of the generated internal waves, and the designed working conditions, and to design and fabricate a navigation body model that meets the size and functional requirements (Fig. 8.8).

(1) Overall Structural Design

The design of the experimental working conditions determines the functions that the navigation body needs to achieve. First, the working condition settings of the navigation body are divided into suspended navigation bodies and self-navigating navigation bodies. Combined with the influence law of internal waves on the appendage performance of the navigation body described above, it is necessary to control the position, speed, and rudder angle of the navigation body. Therefore, the navigation body model needs to achieve four functions: hovering, floating and diving control,

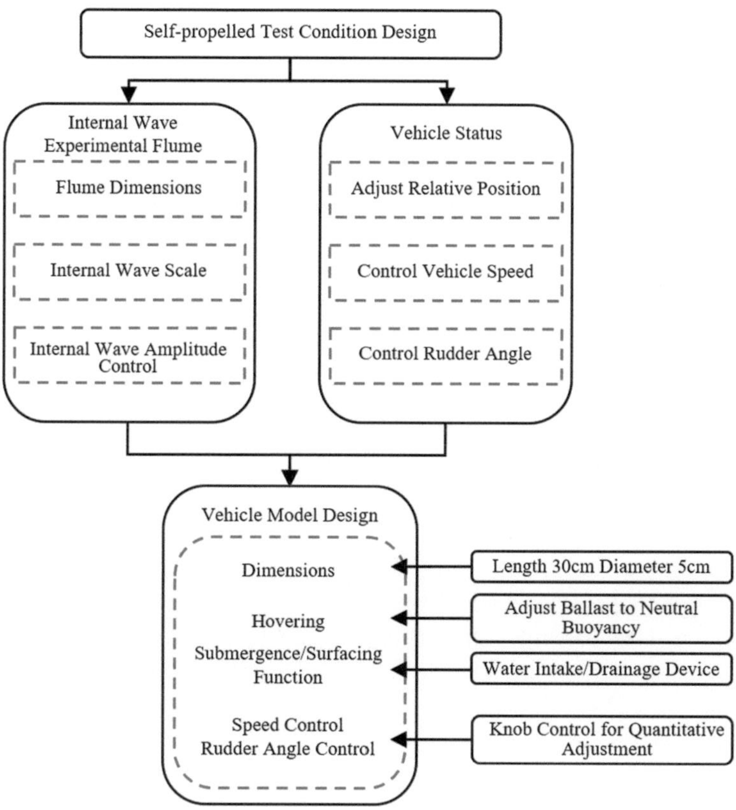

Fig. 8.8 Design and fabrication plan of the navigation body

speed control, and rudder angle control. Among them, the hovering function is achieved by adjusting the counterweight of the navigation body model to adjust the navigation body to zero buoyancy; the floating and diving control is achieved by adding a water suction and drainage device in the cabin; in order to achieve quantitative control of speed and rudder angle, a six-channel remote controller is selected, and the propeller speed adjustment and rudder angle rotation adjustment are connected to the step control knob for control, thereby achieving quantitative adjustment and laying the foundation for subsequent changes in the state of the navigation body. The navigator adopts FM control signal remote control equipment, with a theoretical remote control distance of 500 m on the water surface and a maximum theoretical diving depth of 5 m (Fig. 8.9).

The waterproof design adopts a centralized waterproofing scheme using sealed pressure-resistant waterproof compartments, as shown in Fig. 8.10. The hull, power section, and other connecting components do not require additional waterproofing measures. The brushless motor of the propeller uses industrial-grade waterproof components with strong independent waterproofing capabilities. The steering gear undergoes separate waterproofing treatment: the electronic compartment is coated with conformal paint and filled with vulcanized adhesive, while the mechanical compartment is filled with solid grease. Both the electronic system and power module are centrally sealed within the pressure-resistant waterproof compartment. The waterproof compartment is made of cylindrical acrylic material, ensuring internal visibility as well as pressure resistance and sealing. Both ends are equipped with machined

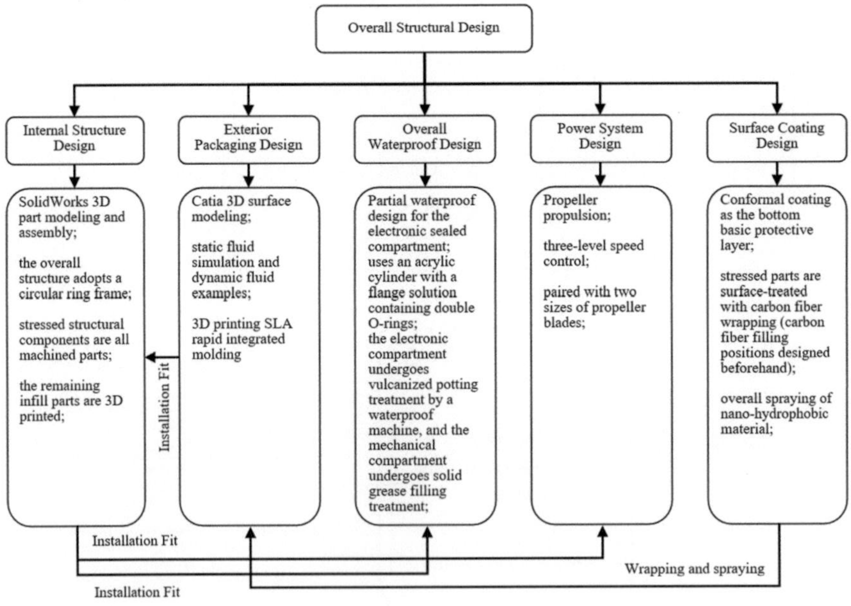

Fig. 8.9 Flow chart of overall structure design

Fig. 8.10 Cylindrical waterproof compartment

aluminum alloy flanges with O-ring grooves, strictly complying with national waterproofing standards. A double-layer O-ring is used to integrate the cylinder, flanges, and hatch into a fully sealed waterproof compartment.

(2) Vehicle Shape

To better simulate real-world conditions, the vehicle adopts a proportionally scaled-down SUBOFF-AFF8 fully appended submarine model. Due to the limited size of the internal solitary wave test tank, the working area is only 1.5 m long, and the maximum amplitude of the generated internal solitary wave is approximately 15 cm. Therefore, the vehicle's size must be minimized as much as possible. Considering the functional requirements, the final dimensions of the vehicle model are set at length $L = 311$ mm, diameter $D = 50$ mm, and height $H = 71$ mm, as shown in Fig. 8.11. The upper and lower surfaces of the hull feature hollowed-out perforations to improve water intake and drainage efficiency.

(3) Functional Implementation

Depth Suspension: Small rectangular counterweights are evenly distributed inside the vehicle's hull. Based on the desired suspension depth, the vehicle is adjusted to neutral buoyancy at that position.

Diving Control: A water intake and drainage device is installed inside the waterproof compartment, consisting of a 1S brushed ESC paired with an N20 lead screw reduction motor water tank. The water tank shell is made from syringe casings, as shown in Fig. 8.12.

Speed Control: A 1S brushless ESC is paired with a waterproof brushless motor. Through a 4-channel mini FM receiver, the ESC is connected to the step-control knob on the remote controller, enabling quantitative speed (propeller RPM) control. Three speed levels are preset, and two propeller sizes are provided, theoretically allowing six-speed control.

Rudder Angle Control: A 1S brushless ESC is paired with a mini servo. Similar to the speed control method, the ESC is connected to the step-control knob on the remote controller via a 4-channel mini FM receiver, with the maximum rudder angle set to 20°.

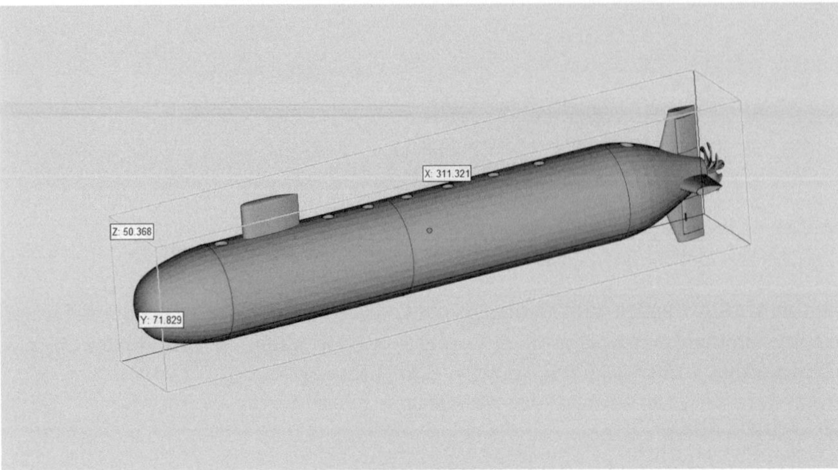

(a) Model Diagram

(b) Physical Diagram

Fig. 8.11 Vehicle for self-propelled experiments in internal wave environments

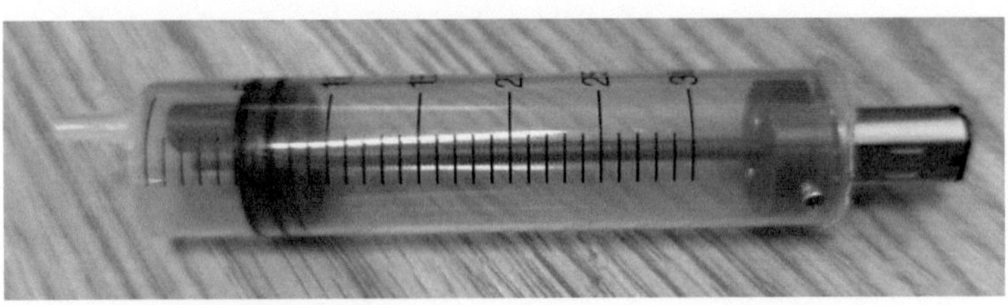

Fig. 8.12 Vehicle water intake and drainage device

8.2.3 Data Acquisition and Post-processing Methods

Wave-current visualization serves as an intuitive observation method in physical model experiments. Using visualization techniques, the propagation and evolution of internal solitary waves and their interaction with structures can be qualitatively analyzed. In the experiment, a conductivity probe array was employed to monitor the fluid density, obtaining the densities of the upper and lower fluid layers. A high-frame-rate industrial camera was used to record the detailed process of internal solitary wave propagation, evolution, and interaction with structures, as well as the motion response of the vehicle under different navigation states in the presence of internal waves. Additionally, transparent scale paper was affixed outside the internal wave flume to determine the vehicle's position using coordinates.

For post-processing, the pixel coordinates of the images were combined with the coordinates from the transparent scale paper. Ten images were captured at equal intervals within 1 s (i.e., one image every 0.1 s) to obtain the centroid coordinates of the vehicle at each 0.1 s interval. This allowed for the derivation of the centroid's motion trajectory and comparative analysis. The results were then compared with numerical simulations to summarize the influence of different states on the vehicle's motion under internal waves and analyze the reasons for changes in the vehicle's motion state.

8.3 Experiments on the Interaction Between Internal Solitary Waves and Suspended Vehicles

Based on the experimental design in Sect. 8.2, this section conducts self-propulsion experiments of underwater vehicles in an internal solitary wave environment according to the working conditions. Using a suspended vehicle within the wave amplitude range as the baseline condition, the motion response of the suspended vehicle under internal waves is explored. The trajectory of the vehicle's center of gravity is recorded and comparatively analyzed. A post-processing and analysis method for the self-propulsion test results is established. Parameters such as the vehicle's position and internal solitary wave amplitude are sequentially varied to investigate the motion response of the vehicle under different working conditions under internal solitary waves. The influence laws of the motion response of suspended vehicles in an internal solitary wave environment are summarized, providing a reference for submerged bodies encountering internal waves in real marine environments.

In the internal solitary wave flume, a strongly stratified two-layer fluid is prepared as the background environment for internal solitary waves. The total fluid height is 50 cm, with the upper layer fluid density set to $\rho_1 = 1000\,\text{kg/m}^3$ (i.e., fresh water) and a height of $h_1 = 10\,\text{cm}$. The lower layer fluid density is set to $\rho_2 = 1020\,\text{kg/m}^3$ (saltwater), with a height of $h_2 = 10\,\text{cm}$. The thickness ratio of the upper and lower fluid layers is $h_1/h_2 = 1/4$. Internal solitary waves are generated using the gravity collapse method. For the flume used in this experiment, the amplitude of the generated internal waves is approximately half the collapse height, with a maximum achievable amplitude of 17.5 cm.

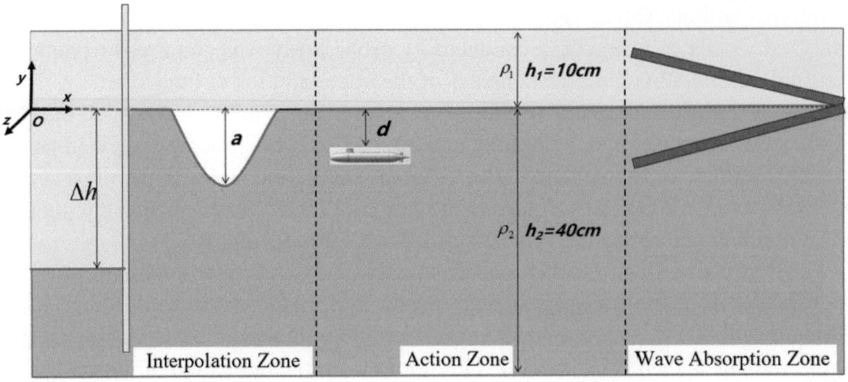

Fig. 8.13 Experimental schematic

The experimental schematic is shown in Fig. 8.13. A coordinate system is established in the internal solitary wave flume as illustrated. The origin is located at the leftmost end of the flume, 10 cm below the free surface—i.e., at the interface between the two fluid layers. The positive x-axis extends from left to right along the flume, representing the propagation direction of the internal solitary wave. The positive y-axis runs vertically upward along the flume depth, and the positive z-axis points toward the front wall of the flume. Δh denotes the collapse height in the wave generation zone. Generally, $a \approx 0.5\Delta h$, though slight variations may occur due to fluid mixing at the stratified interface. d represents the submerged depth of the vehicle. When the vehicle is above the interface, $d > 0$; when at the interface; and when in the lower fluid layer, $d < 0$. In this study, only the motion response of the vehicle in the xy-plane is considered.

8.3.1 *Influence of Submerged Depth*

Since internal solitary waves induce a clockwise-rotating velocity field in the flow, the motion response of the vehicle varies with its submerged depth under the influence of internal solitary waves. This section investigates the effect of submerged depth on the motion response of a suspended vehicle in an internal solitary wave environment. During the experiments, the wave amplitude (i.e., the gravitational collapse height) and the vehicle speed were kept constant (speed = 0). The submerged depth was adjusted solely by modifying the vehicle's buoyancy and ballast. The specific test conditions are listed in Table 8.1. The initial positions for the three test conditions are illustrated in Fig. 8.14. Post-processing of the experimental results yielded the trajectory of the vehicle's center of mass, the pitch angle variation curve, and the velocity variation curve.

Table 8.1 Test conditions for suspended vehicles at different submerged depths

Case	Submerged depth d (cm)	Wave amplitude a (cm)	Speed V (m/s)	Pitch angle ∂
Case 1	+3	10	0	0°
Case 2	0			
Case 3	−7			

Figure 8.15 presents the motion trajectories and pitch angle variations of vehicles at different submerged depths encountering internal waves. During the experiments, due to the vehicle's six degrees of freedom, slight horizontal displacement was unavoidable. To eliminate errors caused by initial position differences, relative coordinates were used for comparing the trajectories of the center of mass across conditions—i.e., the initial coordinates of the vehicle's center of mass were set to (0, 0) for each condition, and the subsequent motion was recorded relative to this starting point. Coordinates are given in pixels, with a distance of 0.7813 mm between two adjacent pixels. For the pitch angle variation curves, the horizontal axis represents time in units of 1/10 s. Since the interaction durations between the vehicle and internal waves differed across conditions, no time normalization was applied to better reflect the actual scenario.

Figure 8.14a shows the centroid motion trajectories of underwater vehicles at different depths encountering internal waves. It can be observed that there are certain differences in the trajectories of vehicles at different depths. The commonality is that in all three working conditions, the vehicles move vertically along the surface of the internal solitary wave, first descending and then ascending after reaching the wave trough. The vehicle at the interface has the largest vertical movement range, reaching 120 unit lengths (approximately 9.4 cm), which is basically consistent with the amplitude of the internal solitary wave. The vehicle in the lower fluid layer has the smallest vertical movement range of only 80 unit lengths, about 0.6 times the wave amplitude. The horizontal movements of vehicles at different depths vary significantly. The upper-layer vehicle always moves along the propagation direction of the internal wave, with a movement range reaching 800 unit lengths. This is because the clockwise velocity field induced by the internal solitary wave in the upper layer flows along the positive x-direction with relatively large amplitude. The vehicle at the interface has a smaller horizontal movement range due to the weaker horizontal velocity in the interfacial flow field. It first moves along the positive x-direction, then reverses near the wave trough, and moves positively again after passing through the trough, resulting in intersecting centroid trajectories that closely match the velocity field in the flow. The lower-layer vehicle, when encountering the internal solitary wave, is always subjected to horizontal velocities along the negative x-direction, thus moving consistently in the negative direction. This demonstrates that the velocity field induced by internal solitary waves is the primary factor influencing the motion response of suspended vehicles.

Figure 8.15b shows the pitch angle variation curves of underwater vehicles at different depths encountering internal waves, where the clockwise direction is defined

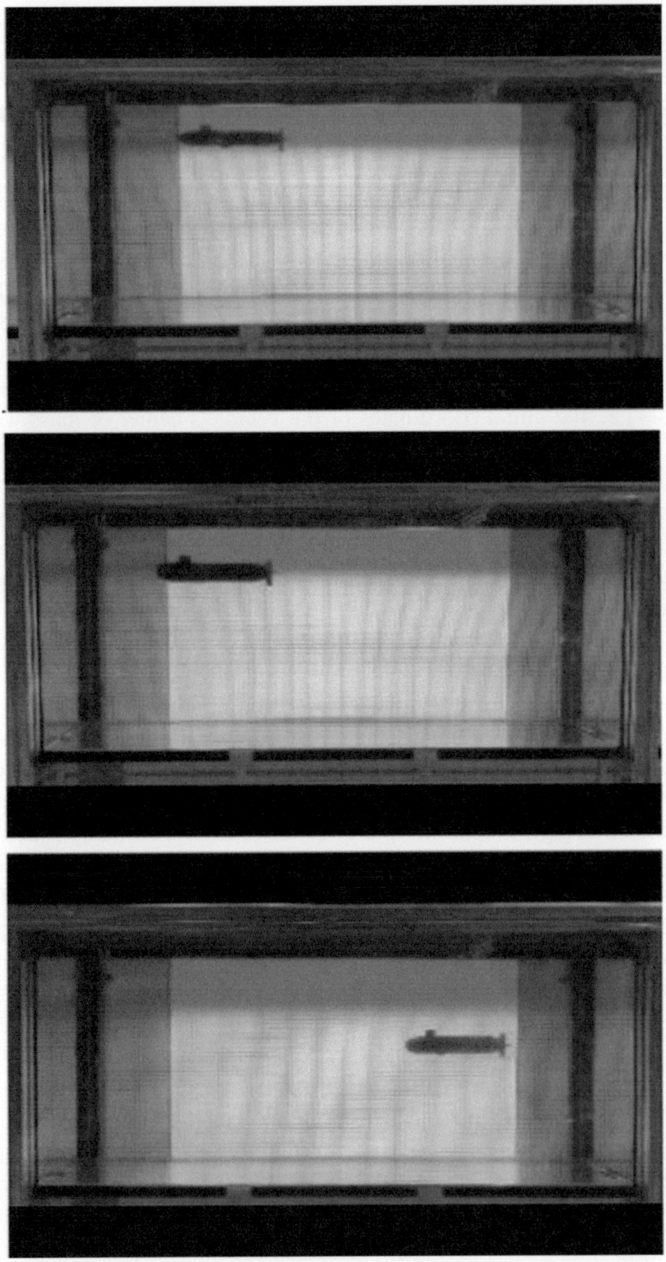

Fig. 8.14 Vehicles at different submerged depths

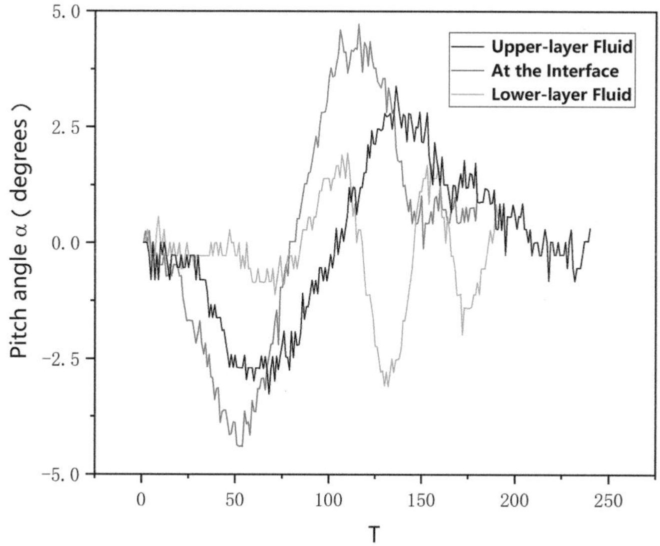

(a) Centroid motion trajectory

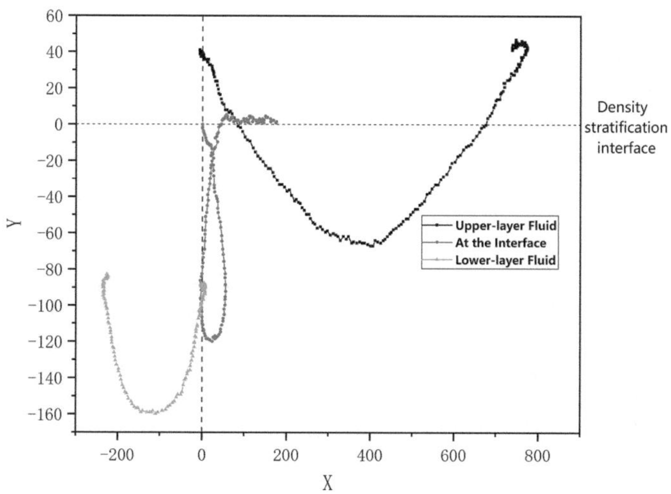

(b) Pitch angle variation curve

Fig. 8.15 Motion responses of underwater vehicles at different depths encountering internal waves

as positive—meaning the pitch angle increases when the vehicle raises its head and decreases when it lowers its head. From the figure, it can be observed that the pitch angle variation trends of vehicles at different depths encountering internal solitary waves are similar. In particular, the vehicles located in the upper fluid layer and at the interface exhibit completely identical pitch angle trends: they first lower their heads when approaching the right wave surface, then raise their heads along the left wave surface after passing the wave trough. The pitch angle variation aligns with the wave surface shape, indicating that the vehicle moves along the wave surface. Consequently, the amplitude of pitch angle variation is larger for the vehicle at the interface. For the vehicle suspended in the lower fluid layer, the trend is more complex because its initial position is already below the interface. When approaching the right wave surface, it moves further downward, so no significant head-lowering trend is observed. Immediately after passing the wave trough, due to its proximity to the left wave surface, a slight head-raising trend occurs. When the vehicle's tail moves below the left wave surface, the upward velocity acting on the tail causes a noticeable head-lowering trend, which gradually subsides. Notably, as the internal wave moves away, the vehicle exhibits another head-lowering trend. This double head-lowering phenomenon does not occur for vehicles at other positions, as shown in Fig. 8.16, indicating that the impact of internal solitary waves on the vehicle in the lower layer is more complex. This also suggests that when the vehicle is in the lower fluid layer, it is more prone to "falling deep", requiring special attention.

To further investigate the influence of internal solitary waves on the motion states of underwater vehicles at different depths, the velocity variation curves of the vehicles were obtained through post-processing of experimental results, as shown in Fig. 8.17.

From the figure, it can be observed that the velocity variations of vehicles at different depths are consistent with the trends of their centroid trajectories. The horizontal velocity of the vehicle in the upper fluid layer always aligns with the positive x-direction, while that of the vehicle in the lower fluid layer always aligns with the negative x-direction, both matching the surrounding fluid velocity directions. At the wave trough position, the velocity reaches its maximum. For the vehicle at the interface, since the horizontal velocity of the flow field is minimal here, the vehicle's velocity is also the smallest. Under all three conditions, the vertical velocity of the vehicles first follows the negative y-direction and then switches to the positive y-direction after passing the wave trough. The vertical velocity is largest at the interface, so the vehicle in Case 2 achieves the greatest vertical velocity.

8.3.2 Influence of Wave Amplitude

As one of the important parameters of internal solitary waves, wave amplitude also affects the motion state of navigation bodies. Therefore, this section focuses on studying the influence of wave amplitude on the motion response of suspended bodies in internal solitary wave environments.

During the experiment, the navigation body was always located at the interface between upper and lower layer fluids, maintaining zero navigation speed. The wave amplitude was adjusted solely by changing the collapse height in the wave-making zone, with three working conditions having collapse heights of 15 cm, 20 cm, and 30 cm respectively. Specific working condition settings are shown in Table 8.2.

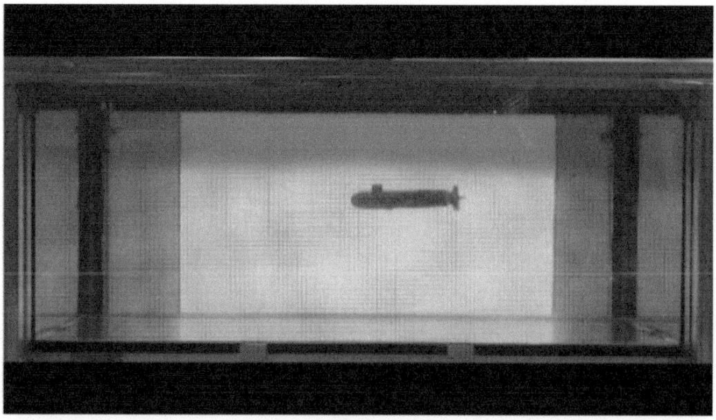

(a) At T=126

(b) At T=167

Fig. 8.16 Head-lowering phenomenon of the vehicle in the lower fluid layer

Through post-processing of experimental results, the motion trajectory of the navigation body's center of mass, pitch angle variation curve, and velocity variation curve were obtained.

Figure 8.17 shows the motion trajectory and pitch angle variation of the navigation body encountering internal waves under different wave amplitude conditions. The coordinate system settings and post-processing methods in the figure remain consistent with those in Sect. 3.1.

Figure 8.18a shows the centroid motion trajectories of the vehicle when encountering internal waves under different wave amplitude conditions. It can be found that the motion trajectories of the vehicle under the three conditions have basically the same changing trend: it first moves downward to the right, and then moves upward to the right after passing through the position of the wave trough. The motion trajectory is consistent with the shape of the internal solitary wave surface, and only the motion range is different. The larger the amplitude of the internal solitary wave is, the greater

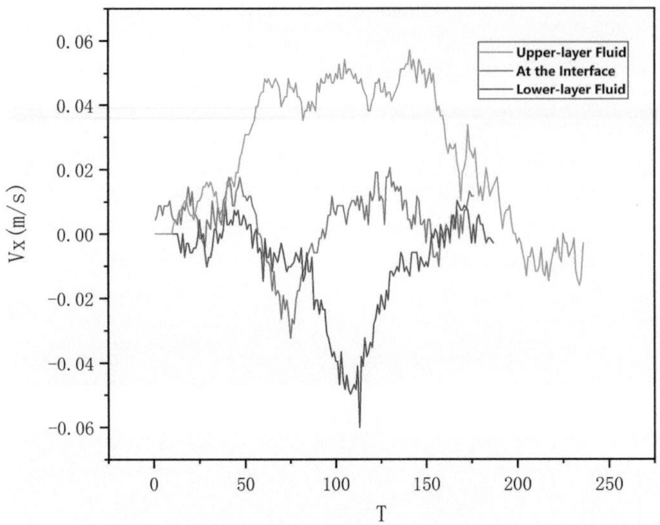

(a) Horizontal velocity

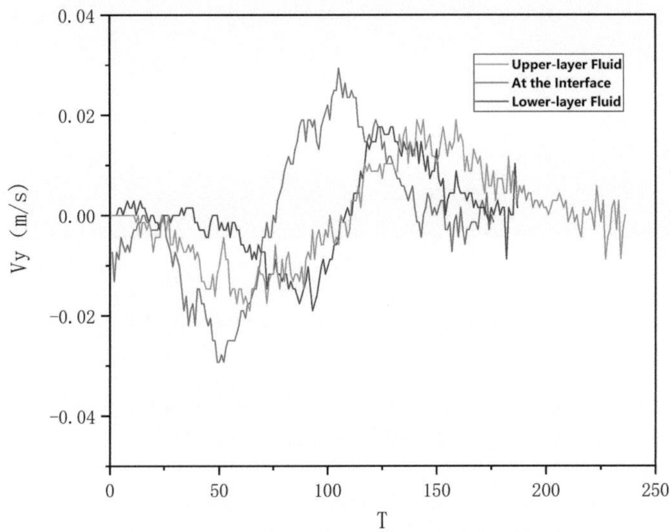

(b) Vertical velocity

Fig. 8.17 Velocity variation curves of underwater vehicles at different depths encountering internal waves

Table 8.2 Working condition settings for suspended submersible under different internal solitary wave amplitudes

Case	Submerged depth d (cm)	Wave amplitude a (cm)	Speed V (m/s)	Pitch angle ∂
Case 4	0	7.5	0	0°
Case 5		10		
Case 6		15		

the horizontal and vertical motion distances of the vehicle are. In Case 4, the vertical motion distance of the vehicle is approximately 78 unit lengths (6.1 cm); in Case 5, it is 120 unit lengths (9.4 cm); and in Case 6, it is 155 unit lengths (approximately 12.1 cm). The vertical motion distances in the three conditions are all close to the corresponding wave amplitude heights, which further indicates that the suspended vehicle located at the interface moves along the internal solitary wave surface and will not enter the lower fluid. In the horizontal direction, only in Case 3 does the motion distance of the vehicle in the negative direction of the x-axis exceed the initial position, but it still finally returns to the right side of the initial position, indicating that the suspended vehicle located at the interface, under the action of internal solitary waves, will always finally move to the front of the initial position along the wave propagation direction.

Figure 8.18b shows the variation curves of the pitch angle of the vehicle encountering internal waves under different wave amplitude conditions. It can be observed from the figure that under different wave amplitude environments, the variation trend of the pitch angle of the vehicle at the same position is completely consistent. The vehicle first pitches downward along the tangent direction of the internal solitary wave surface. In this stage, the larger the wave amplitude, the greater the downward pitch angle of the suspended vehicle, and the more pronounced the downward trend, making it more dangerous. As the vehicle approaches the wave trough, the pitch angle begins to decrease, and after passing the trough, an upward pitching trend appears. In this stage, a larger wave amplitude does not result in a greater upward pitch angle. Instead, under large wave amplitude conditions, the upward pitch angle of the vehicle is slightly smaller than in the other two cases. The reason for this may be that under large wave amplitude conditions, the vehicle's downward pitch angle is larger, requiring a greater restoring force and a longer recovery period, which is insufficient to produce a large upward pitch angle within the short encounter time with the internal solitary wave. Notably, as the wave amplitude increases, the fluctuation of the pitch angle also becomes larger, indicating that the larger the wave amplitude, the more unstable the vehicle's state, requiring greater caution.

Similarly, post-processing of the experimental results also yielded the velocity variation curves of the vehicle under different wave amplitude conditions, as shown in Fig. 8.19.

From the figure, it can be observed that the variation trends of the horizontal and vertical velocities of the vehicle under the three different wave amplitude conditions are completely consistent. The larger the wave amplitude, the greater the

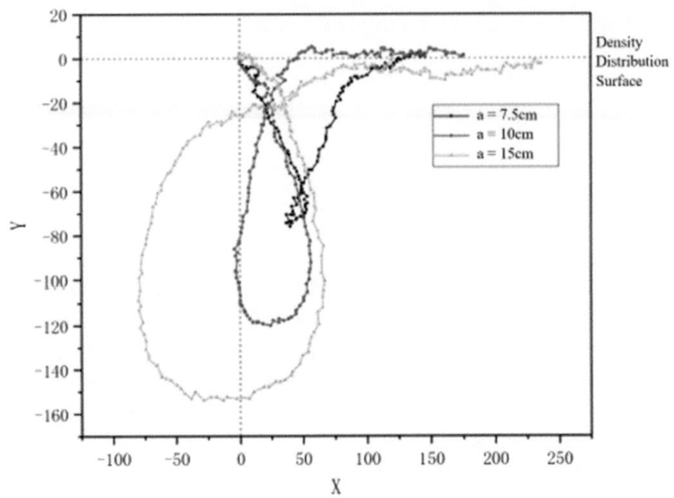

(a) Center of Mass Motion Trajectory

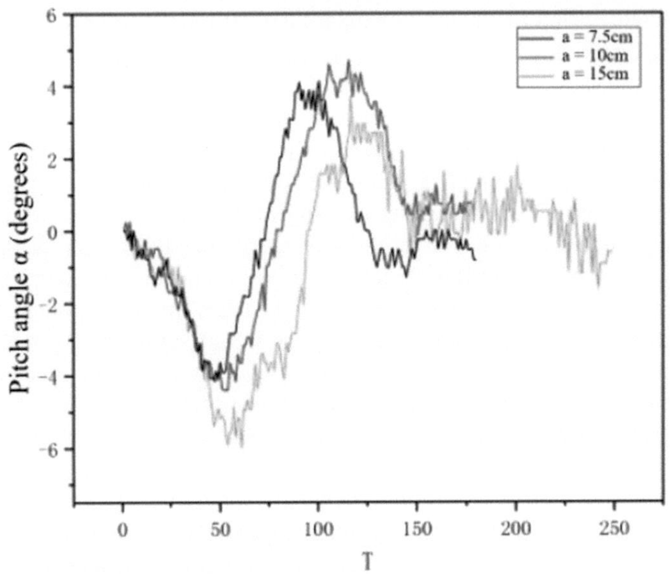

(b)Pitch Angle Variation Curve

Fig. 8.18 Motion response of navigation body encountering internal waves under different wave amplitude conditions

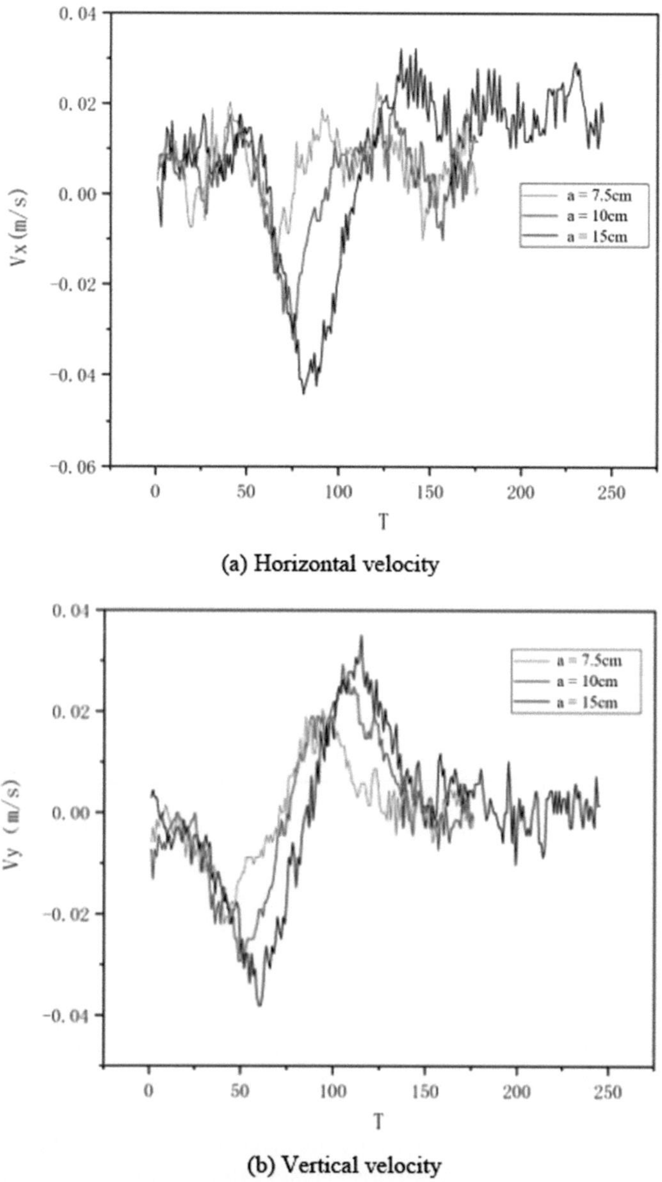

Fig. 8.19 Velocity variation curves of the vehicle under different wave amplitude conditions

velocity generated by the suspended vehicle, which aligns with the variation trend of the obtained centroid motion trajectory. This further confirms the aforementioned phenomena and the analysis of their causes.

8.4 Experiments on the Interaction Between Internal Solitary Waves and Self-propelled Vehicles

Based on the experimental study of internal solitary waves and suspended bodies in Sect. 8.3, this section conducts experiments on the motion response of self-propelled vehicles in an internal wave environment according to the working conditions. Taking the self-propelled vehicle traversing the internal wave as the baseline condition, the motion response of the self-propelled vehicle under the influence of internal waves is explored. Furthermore, based on the influence laws of internal waves on the hydrodynamic performance of vehicle appendages obtained in Chaps. 2 and 3, parameters such as internal solitary wave amplitude, speed (propeller rotation speed), rudder angle, and submerged depth are sequentially varied to investigate the motion response of self-propelled vehicles under different navigation states in the presence of internal waves. The influence laws of navigation parameters on the motion response of self-propelled vehicles in an internal wave environment are summarized, providing a reference for the maneuvering strategies of self-propelled submerged bodies encountering internal waves in real marine environments.

The experimental setup of the internal solitary wave tank in this section is identical to that in Sect. 8.3. In all working conditions, the vehicles are self-propelled with speed. The definitions and representations of the parameters involved in the experiments are also exactly the same.

8.4.1 Influence of Wave Amplitude

Since the self-propelled vehicle has its own speed, its motion response characteristics will differ even under the same internal solitary wave conditions. This section investigates the influence of wave amplitude on the motion response of the self-propelled vehicle in an internal solitary wave environment. Compared to a suspended body, the self-propelled vehicle is subject to more complex effects from internal waves. During the experiment, the vehicle was placed in both the upper fluid layer and at the density interface, maintaining the same initial speed and zero initial pitch angle. For each of the two submergence depths, three internal solitary wave amplitude conditions were tested, with collapse heights of 15 cm, 20 cm, and 30 cm, corresponding to wave amplitudes of approximately 7.5 cm, 10 cm, and 15 cm, respectively. The specific experimental conditions are listed in Table 8.3.

Table 8.3 Experimental conditions for self-propelled vehicles under different internal solitary wave amplitudes

Case	Submergence depth d (cm)	Wave amplitude a (cm)	Speed V (m/s)	Pitch angle ∂
Case 7	4	7.5	0.1	0°
Case 8		10		
Case 9		15		
Case 10	0	7.5	0.08	
Case 11		10		
Case 12		15		

(1) Interaction between internal solitary waves and self-propelled vehicles in the upper fluid layer

This subsection focuses on Cases 7–9, where internal solitary waves of three different amplitudes were generated for self-propelled vehicles at the same submergence depth in the upper fluid layer. The changes in motion trajectory and state under the influence of internal solitary waves were monitored. Post-processing techniques were used to obtain the vehicle's centroid trajectory, pitch angle variation, and horizontal and vertical velocity change curves. Figure 8.20 illustrates the motion trajectory and pitch angle variation of the self-propelled vehicle in the upper fluid layer encountering internal waves under different wave amplitude conditions. The coordinate system setup and post-processing methods remain consistent with those in Sect. 8.3.

Figure 8.20a shows the centroid motion trajectories of a self-propelled vehicle encountering internal waves in the upper fluid layer under different wave amplitude conditions. It can be observed that the motion trajectories under all three wave amplitude conditions follow nearly identical trends: first moving downward along the right wave surface of the internal solitary wave, reaching the wave trough, then ascending along the left wave surface, with the final position always above the initial starting point. This phenomenon occurs because the vehicle maintains a positive pitch angle when moving along the left wave surface, as evidenced in Fig. 8.20b. Due to the vehicle's inherent velocity, it continues to pitch upward and move leftward for some distance after passing through the solitary wave. Interestingly, the figure clearly shows that as wave amplitude increases, the vertical displacement of the vehicle decreases—a result opposite to that observed in experiments with suspended vehicles. Analysis of pitch angle curves and experimental records reveals that for self-propelled vehicles, their forward speed prevents them from strictly following the wave surface. When the bow approaches the right wave surface, the downward velocity field induces a slight bow-down tendency (Fig. 8.21a). After the bow passes through, the stern enters this region and experiences downward force, gradually creating a bow-up tendency (Fig. 8.21b). In the upper fluid layer, with rightward horizontal velocity, the increased flow-facing area due to pitching up further amplifies the bow-up angle. The vehicle's own speed then causes it to move along the pitching-up direction rather than continuing downward along the wave surface. Larger wave

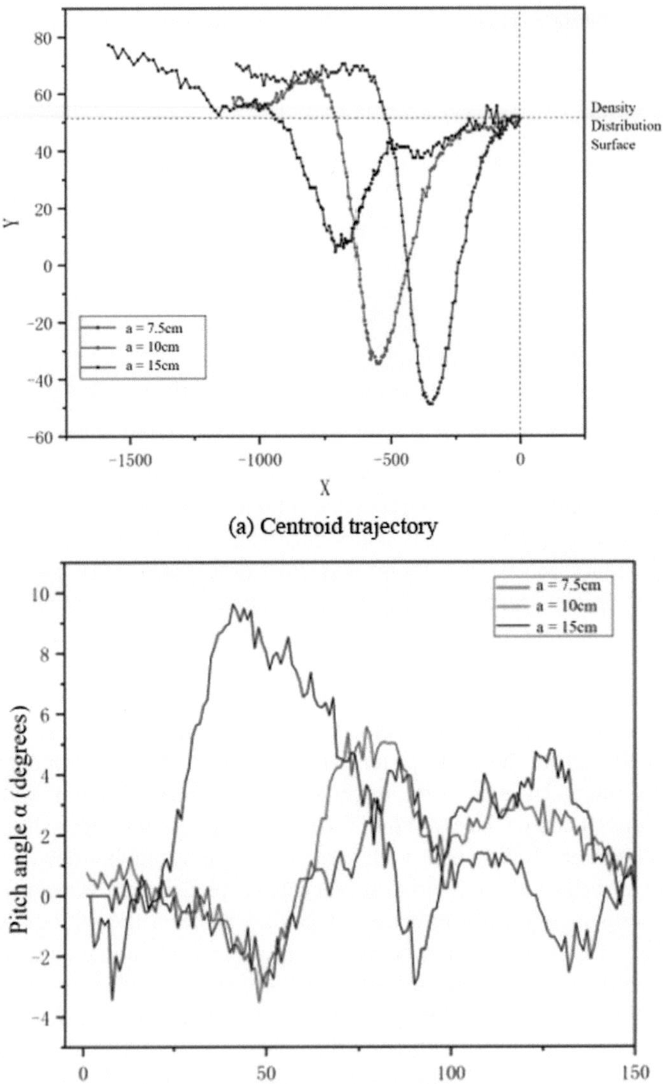

(a) Centroid trajectory

(b) Pitch angle variation

Fig. 8.20 Motion response of self-propelled vehicles in the upper fluid layer encountering internal waves under different wave amplitude conditions

amplitudes induce greater horizontal velocities in the upper layer, resulting in larger pitch-up angles and consequently smaller downward vertical displacements. This conclusion is verified by the pitch angle curves in Fig. 8.20b, which show consistent initial bow-down angles across different wave amplitudes but increasing bow-up angles with larger waves—reaching nearly 10° at 15 cm wave amplitude. The pitch angle variation pattern remains fundamentally similar (first bow-down then bow-up), though notably, a secondary small bow-down occurs after the vehicle passes the left wave surface (Fig. 8.22), caused by upward fluid velocity on the left side of the solitary wave field lifting the stern—a phenomenon requiring careful consideration in real-world applications.

From the above results, it can be observed that for a self-propelled vehicle moving in the upper fluid layer, as the wave amplitude increases, the diving depth of the vehicle decreases. However, the variation in pitch angle becomes more pronounced, with increased amplitude, further intensifying its impact on the vehicle's attitude. This poses a significant threat to the safe navigation of the vehicle.

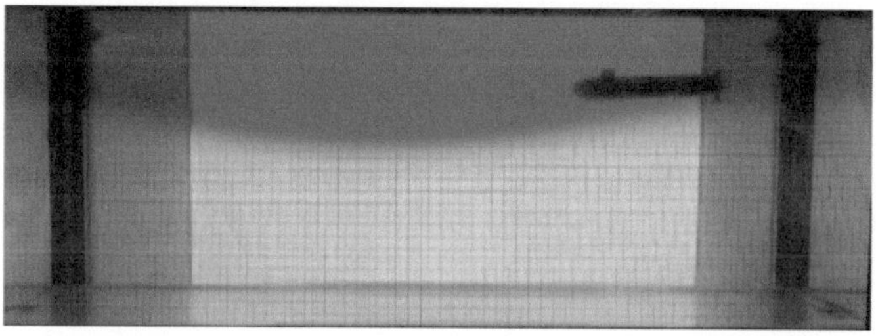

(a) Located above the right wave surface

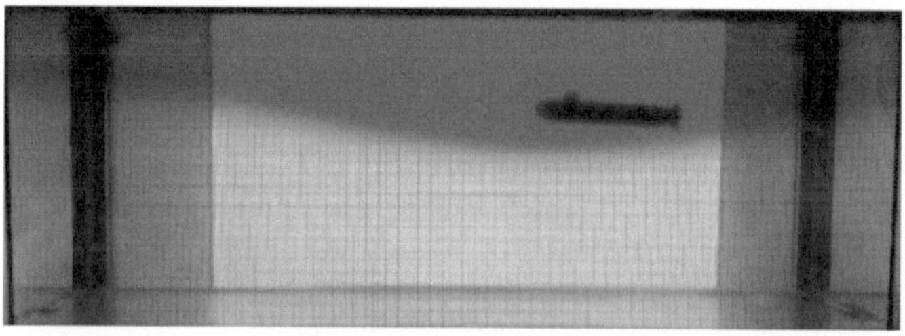

(b)Approaching the wave trough

Fig. 8.21 Motion posture of the vehicle encountering internal waves in Case 8

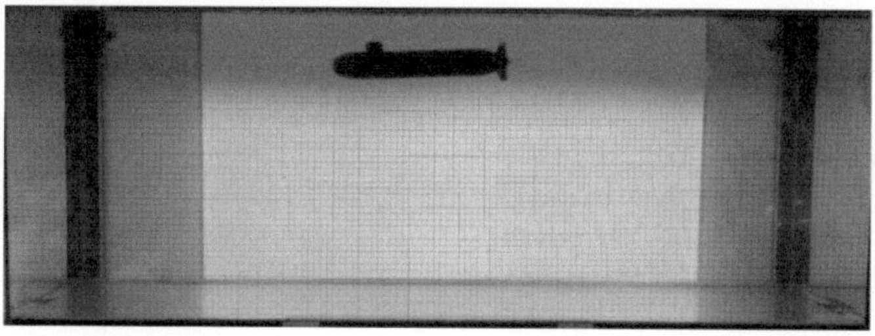

Fig. 8.22 Motion posture of the vehicle after moving away from the internal solitary wave in Case 7

To gain a deeper understanding of the influence of wave amplitude on the self-propelled body in the upper layer, the horizontal and vertical velocity variation curves of the vehicle were obtained through post-processing of experimental results, as shown in Fig. 8.23. It can be observed that the effects of wave amplitude on the horizontal and vertical velocities of the self-propelled body in the upper fluid are opposite. For horizontal velocity, the greater the wave amplitude, the more significant the speed reduction; for vertical velocity, the greater the wave amplitude, the smaller the speed increase. The reason for the horizontal velocity change is that when the vehicle is located in the upper fluid and encounters an internal solitary wave, the surrounding velocity along the positive x-axis opposes the vehicle's motion direction, hindering its movement. Additionally, as discussed in Chap. 3 regarding the interaction between internal solitary waves and propellers, when the propeller is above the wave trough, the thrust decreases substantially. The combined effect of these two factors leads to a reduction in horizontal velocity, with larger wave amplitudes resulting in stronger hindrance. The change in vertical velocity follows the same principle as the previously mentioned decrease in vertical displacement with increasing wave amplitude, and the vertical velocity trend further confirms the variation pattern in vertical displacement.

(2) Interaction between internal solitary waves and the self-propelled body at the interface

This subsection focuses on Cases 10–12, where internal solitary waves of three different amplitudes were generated for the self-propelled body located at the density-stratified interface. The changes in motion trajectory and state under the influence of internal solitary waves were monitored, and post-processing techniques were used to obtain the centroid trajectory, pitch angle variation, and horizontal and vertical velocity curves of the self-propelled body. Figure 8.24 illustrates the motion trajectory and pitch angle changes of the self-propelled body at the fluid interface encountering internal waves under different wave amplitude conditions.

Figure 8.24a shows the centroid motion trajectory of the self-propelled body at the fluid interface encountering internal waves under different wave amplitude conditions. It can be observed that the variation pattern of the motion trajectory for the self-propelled body at the interface is relatively simple. The trends of the motion trajectories under the three wave amplitude conditions are entirely consistent—all moving along the wave surface direction. Moreover, the greater the wave amplitude, the larger the motion range in both horizontal and vertical directions, as an increase in wave amplitude is accompanied by an elongation of the wavelength.

Figure 8.24b shows the variation curves of the pitch angle of the self-propelled vehicle at the fluid interface encountering internal waves under different wave amplitude conditions. It can be observed that the wave amplitude has little effect on the pitch angle variation of the vehicle at this position. Unlike the suspended body, the vehicle exhibits a slight upward tilt at the beginning of its motion. This is likely due to the sudden change in the vehicle's speed from zero, where the thrust generated at the tail causes an inertial upward tilt. Compared to the suspended body, the self-propelled vehicle shows a smaller downward tilt trend on the right wave surface

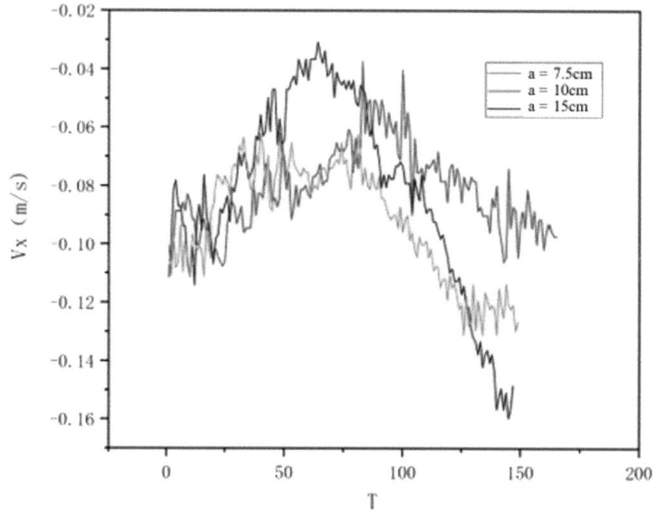

(a)Horizontal velocity

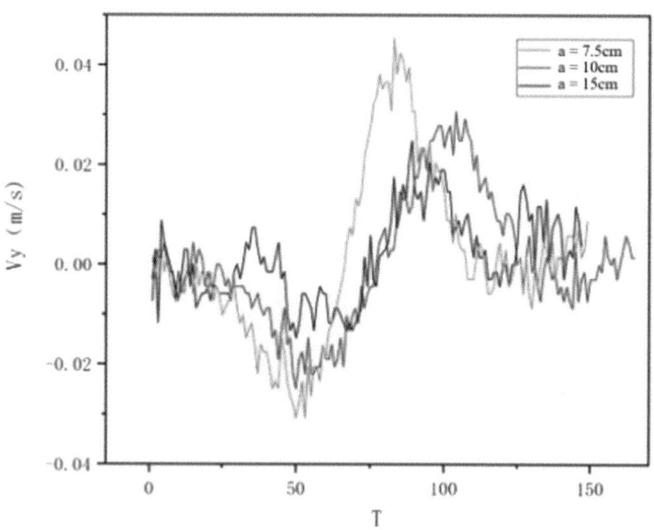

(b)Vertical velocity

Fig. 8.23 Velocity variation curves of the self-propelled body in the upper fluid under different wave amplitudes

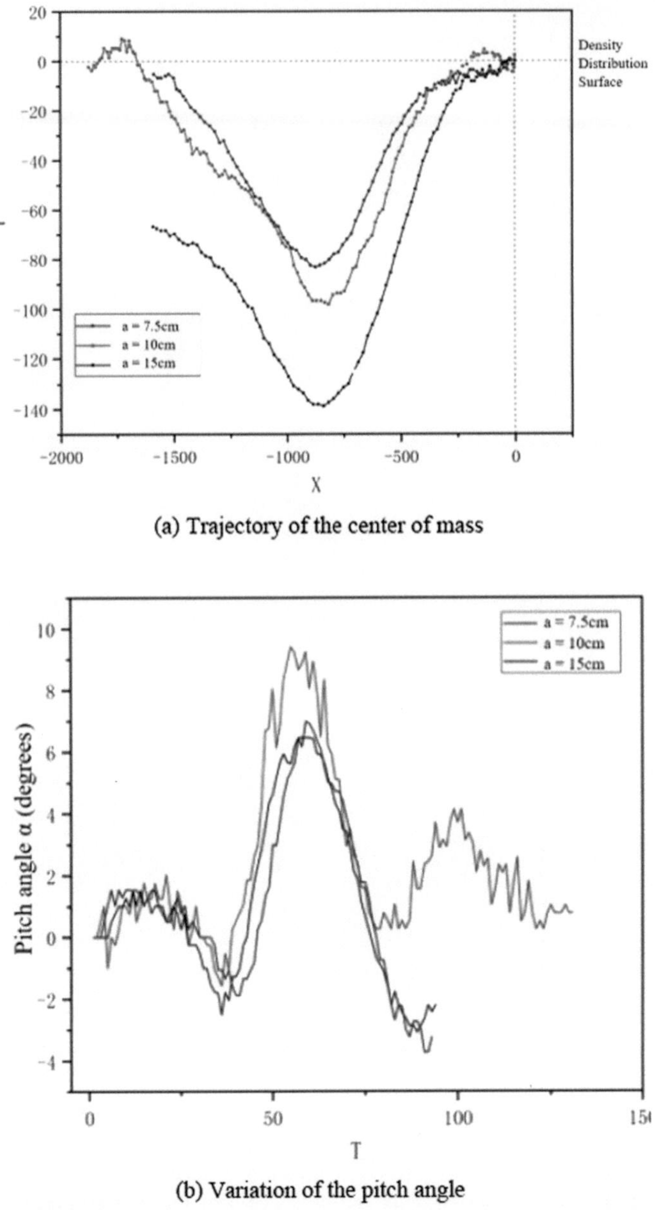

Fig. 8.24 Motion response of the self-propelled vehicle at the interface encountering internal waves under different wave amplitudes

of the internal solitary wave. This is because the vehicle's existing speed prevents it from fully following the wave surface to the trough. Notably, after the vehicle completely passes through the internal solitary wave, it exhibits another downward tilt phenomenon, the reason for which can be referenced in the previous analysis of the motion patterns of self-propelled vehicles in the upper fluid. Specifically, in the case of a 10 cm wave amplitude, the vehicle does not show a downward tilt in the final stage but instead tilts upward. Analysis of the experimental flow field reveals that, under this condition, the vehicle enters the lower fluid near the trough and moves upward to the right after passing through the trough, as shown in Fig. 8.25. Therefore, it is not affected by the left wave surface of the internal wave.

Similarly, post-processing of the experimental results yielded the horizontal and vertical velocity variation curves of the self-propelled vehicle at the interface, as shown in Fig. 8.26. From Fig. 8.23, it can be observed that under different internal solitary wave amplitudes, the trends of horizontal and vertical velocity changes when the vehicle at the interface encounters internal waves are generally consistent, differing only in amplitude. For the horizontal velocity, as the vehicle moves along the wave surface, the velocity continuously increases due to the clockwise velocity field. When the vehicle reaches the wave trough, the horizontal velocity reaches its maximum value, with larger wave amplitudes corresponding to higher maximum horizontal velocities, followed by a gradual recovery. For the vertical velocity, the vehicle primarily moves up and down along the wave surface. An

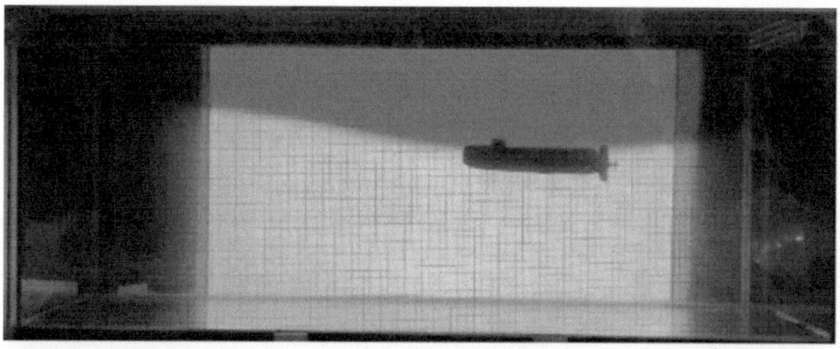

(a) Positioned at the wave trough

(b) Positioned below the left wave surface

Fig. 8.25 Motion posture of the vehicle encountering internal waves in Case 11 increase in wave amplitude enhances the vertical velocity when descending along the right wave surface but has little effect on the ascending velocity along the left wave surface.

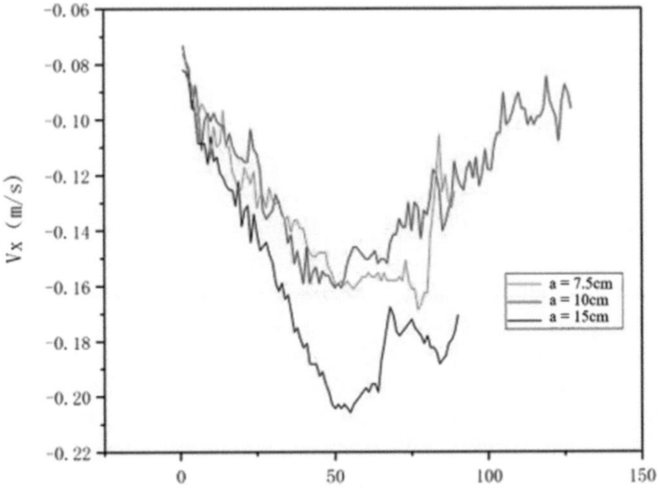

(a) Horizontal velocity

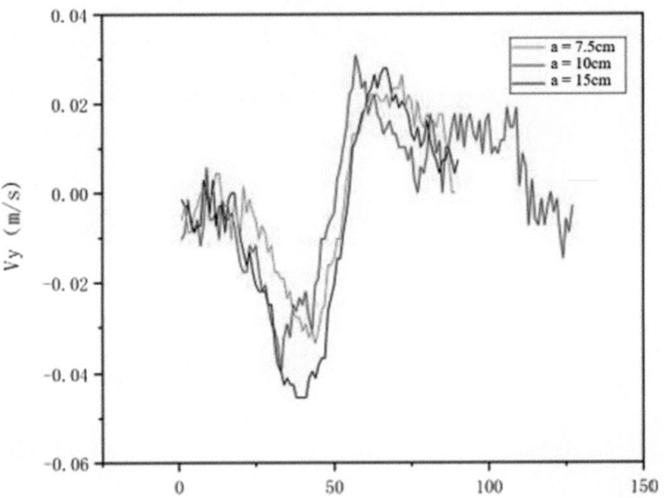

(b) Vertical velocity

Fig. 8.26 Velocity variation curves of the self-propelled vehicle at the interface under different wave amplitudes

Through the above research, it was found that the self-propelled vehicle at the interface exhibits opposite motion patterns compared to those in the upper layer. At the interface, larger wave amplitudes lead to greater motion ranges in both horizontal and vertical directions. Additionally, the attitude (pitch angle) changes of the self-propelled vehicle at the interface are relatively stable and less sensitive to variations in wave amplitude.

8.4.2 *Influence of Navigation Speed*

Navigation speed is one of the important motion parameters of self-propelled vehicles, which also affects their motion state. Therefore, this section studies the influence of vehicle speed on the motion response of self-propelled vehicles in the internal solitary wave environment.

During the experiment, the vehicle was always located at the interface between the upper and lower fluid layers, maintaining a consistent collapse height of 30 cm in the wave-making area (i.e., the amplitude of internal solitary waves was consistently 15 cm). The vehicle speed was adjusted with four different speed conditions set up, as detailed in Table 8.4. Through post-processing of experimental results, the motion trajectory of the vehicle's centroid, pitch angle variation curve, and speed variation curve were obtained.

Figure 8.24 shows the motion trajectory and pitch angle variation of self-propelled vehicles encountering internal waves at the interface under different speed conditions.

From Fig. 8.24a, it can be observed that under different navigation speed conditions, the variation trends of the mass center trajectories of the navigational body are fundamentally consistent, still moving along the surface of the internal solitary wave, albeit with different amplitudes of motion. As the navigation speed decreases, the trajectory of the navigational body exhibits greater deviation after encountering the internal solitary wave. Specifically, at lower speeds (e.g., V = 0.06 m/s and 0.08 m/s), the trajectory shows larger fluctuations with more significant downward vertical displacement, indicating that low-speed navigational bodies are more susceptible to the influence of internal solitary waves. The reason for this can be analyzed in conjunction with the pitch angle variations shown in Fig. 8.23b. The pitch angle variation trends for navigational bodies at different speeds are also fundamentally

Table 8.4 Operating conditions for self-propelled vehicles encountering internal waves at different speeds

Case	Submergence depth d (cm)	Speed V (m/s)	Wave amplitude a (cm)	Pitch angle ∂
Case 13	0	0.12	10	0°
Case 14		0.10		
Case 15		0.08		
Case 16		0.06		

consistent, all showing an initial bow-down tendency, followed by bow-up when approaching the wave trough, and then another bow-down trend after passing the left surface of the internal solitary wave. The figure reveals that as the navigation speed decreases, the amplitude of pitch angle variation increases. At higher speeds, the pitch angle changes are relatively smaller, and the navigational body maintains a more stable attitude. Specifically, at lower speeds (e.g., $V = 0.06$ m/s), the pitch angle variation range is larger, with more pronounced fluctuations from positive to negative angles. As speed increases, the amplitude of both bow-down movements significantly decreases, while the bow-up amplitude after passing the wave trough increases with speed. This indicates that low-speed navigational bodies are more prone to significant attitude changes under the influence of internal solitary waves, whereas high-speed navigational bodies can better resist the hydrodynamic disturbances caused by internal solitary waves. The underlying reason is that low-speed navigational bodies, due to their lower kinetic energy, are more easily affected by the hydrodynamic forces of internal solitary waves, while high-speed navigational bodies, with their higher kinetic energy, can better maintain their original attitude. Precisely because high-speed navigational bodies exhibit reduced bow-down tendencies when approaching internal solitary waves, with periodic changes quickly transitioning to bow-up, the downward vertical movement distance of the navigational body consequently becomes shorter. Additionally, it is particularly noteworthy that when the navigational body is at the interface, after passing the left surface of the internal solitary wave, the upward direction of the surrounding velocity field exerts an upward force on the tail, causing the navigational body to exhibit a bow-down tendency, which poses a risk of sudden depth loss (Figs. 8.27 and 8.28).

To further investigate the mechanism of different speeds on self-propelled bodies at the interface, the horizontal and vertical velocity variation curves of the body were obtained through post-processing, as shown in Fig. 8.29. From the figure, it can be observed that the horizontal velocity trends of bodies at different speeds first increase, then decrease, and subsequently rise again. As the body approaches the right wave surface of the internal solitary wave, it exhibits a bow-down tendency. Simultaneously, the surrounding velocity field points toward the lower left, aligning with the body's motion. Under their combined effect, the body's velocity rapidly increases, reaching its maximum near the wave trough. Afterward, the body begins to pitch up, causing the velocity to decline. Once the body moves past the left wave surface, the pitch angle gradually recovers, and the horizontal velocity steadily increases. The trends and amplitudes of vertical velocity changes for bodies at different speeds are largely consistent, all moving upward and downward along the wave surface. Figure 8.29b reveals that lower speeds result in longer periods of vertical velocity variation, leading to greater vertical displacement—consistent with the vertical displacement results shown in Fig. 8.27a.

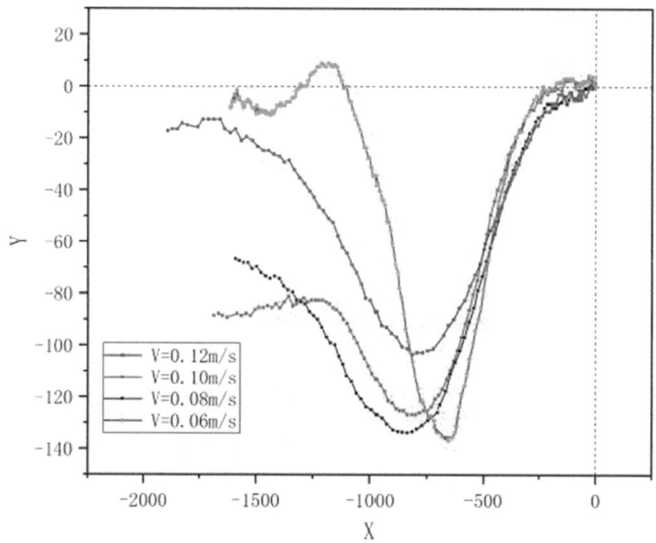

(a) Centroid trajectory

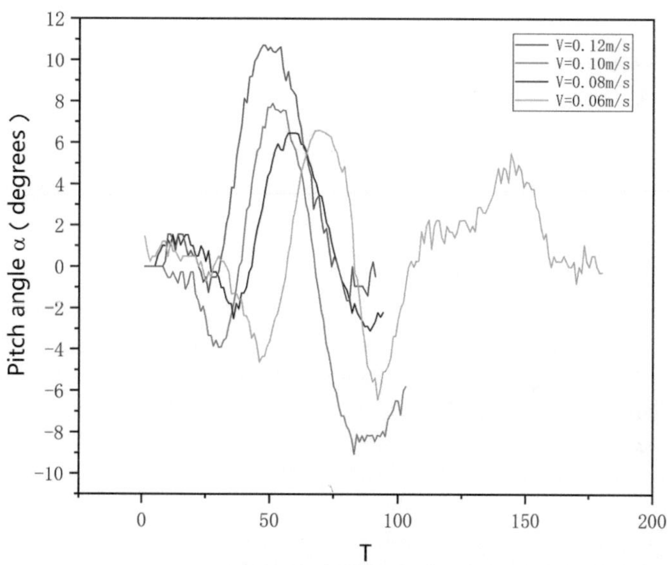

(b) Pitch angle variation

Fig. 8.27 Motion response of vehicles at different speeds encountering internal waves at the interface

(a)V=0.06m/s

(b)V=0.10m/s

Fig. 8.28 Motion posture of navigation body moving away from internal waves at different speeds

8.4.3 Influence of Attitude

In addition to speed, the pitch angle is another crucial parameter affecting the motion of the body. Therefore, this section investigates the influence of the pitch angle on the motion response of self-propelled bodies in an internal solitary wave environment.

During the experiments, the body was placed in both the upper fluid layer and at the density interface, with the collapse height in the wave-making region consistently maintained at 20 cm (i.e., the amplitude of the internal solitary wave was always 10 cm). The pitch angle of the body was adjusted in both the upper fluid layer and at the interface, with three test conditions set for each position. The specific test conditions are listed in Table 8.5. Through post-processing of the experimental results, the motion trajectory of the body's center of mass, pitch angle variation curves, and velocity variation curves were obtained.

(1) Interaction between self-propelled bodies with different pitch angles in the upper fluid layer and internal solitary waves

This subsection focuses on working conditions Cases 17–19, where three navigation postures are set for self-propelled vehicles at the same submerged depth in the upper

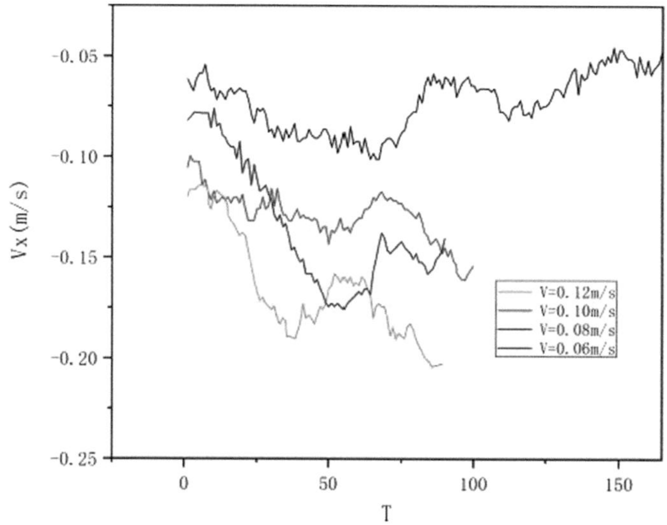

(a) Horizontal velocity

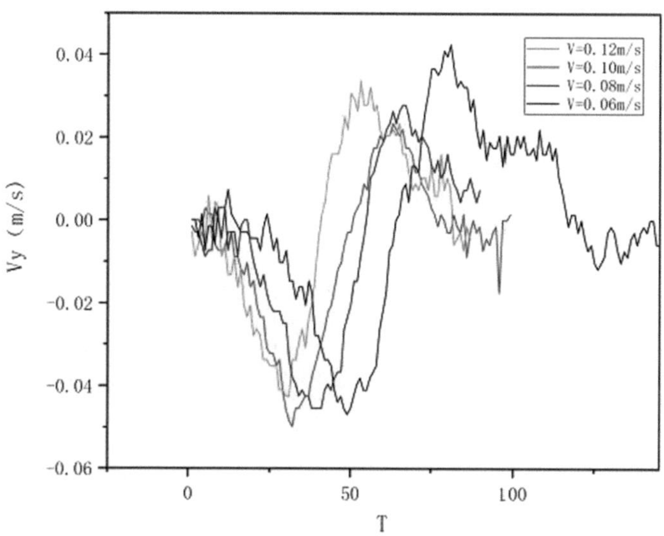

(b) Vertical velocity

Fig. 8.29 Velocity variation curves of self-propelled bodies at different speeds located at the interface

Table 8.5 Test conditions for self-propelled bodies with different pitch angles encountering internal waves

Case	Submergence depth d (cm)	Pitch angle ∂ (attitude)	Wave amplitude a (cm)
Case 17	4	+7° (bow-up) I	10
Case 18		0° (horizontal)	
Case 19		−12° (bow-down)	
Case 20	0	+5° (bow-up)	
Case 21		0° (bow-down)	
Case 22		−14° (bow-down)	

fluid layer, as shown in Fig. 8.30. With consistent internal solitary wave amplitude, the changes in their motion trajectories and states under the influence of internal solitary waves are monitored. Post-processing techniques are used to obtain the centroid trajectory, pitch angle variation, and curves of horizontal and vertical velocity changes of the self-propelled vehicle.

Figure 8.30 illustrates the motion trajectories and pitch angle variations of self-propelled vehicles in the upper fluid layer encountering internal waves under different motion postures.

Figure 8.31a presents the centroid motion trajectories of vehicles at different pitch angles encountering internal solitary waves. It can be observed that the motion trajectories of vehicles at pitch angles of 0° and −12° follow a similar pattern, whereas the motion of the vehicle at +7° pitch angle is entirely different from the other two conditions. The vehicle in a nose-up posture does not move along the wave surface but instead ascends in the direction of its raised nose. In contrast, vehicles in horizontal and nose-down postures move along the wave surface, descending first, passing through the wave trough, and then ascending, with the nose-down vehicle exhibiting a larger vertical motion range.

Figure 8.31b shows the pitch angle variation curves of vehicles under different conditions. From the figure, it can be seen that for the nose-up vehicle, the pitch angle only increases further during the passage through the internal wave, without any decrease or nose-down tendency. Consequently, its motion trajectory does not show a downward trend. This is because the vehicle is located in the upper fluid layer, where the flow velocity is always horizontally rightward, resulting in a continuously applied clockwise moment that increases the pitch angle. The flow velocity reaches its maximum above the wave trough, where the pitch angle also peaks. On the other hand, vehicles in horizontal and nose-down postures still move along the wave surface, initially experiencing a nose-down trend or further increasing their nose-down angle. After passing the wave trough, they are subjected to an upward force, causing the pitch angle to decrease and eventually transition into a nose-up posture.

Through post-processing of experimental results, the horizontal and vertical velocity variation curves of the navigation body were obtained, as shown in Fig. 8.31. From the figure, it can be observed that the horizontal velocities of navigation bodies in all three attitudes decrease as they approach the internal solitary wave. Among

(a)Case 17

(b)Case 18

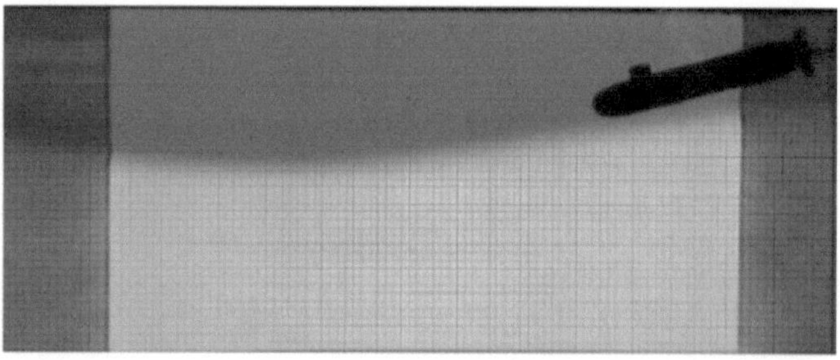

(c)Case 19

Fig. 8.30 Initial postures of vehicles at different pitch angles in the upper fluid layer

them, the initial horizontal and upward-tilted cases show more significant velocity reduction, especially for the horizontal attitude navigation body. Combining with the third conclusion, the reason is considered to be that the propeller of the horizontal attitude navigation body is closest to the trough of the internal solitary wave, thus experiencing the largest thrust reduction. The vertical velocity changes correspond with the motion trajectory patterns: the vertical velocity of the upward-tilted navigation body fluctuates around zero, mostly above zero, while the horizontal and downward-tilted navigation bodies first increase their negative vertical velocity, gradually recover near the wave trough, then increase in the positive direction. The downward-tilted navigation body shows the largest variation range in vertical velocity, indicating this

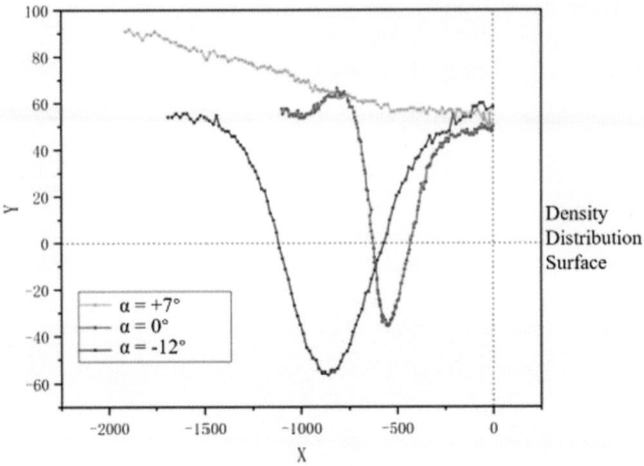

(a) Centroid trajectory

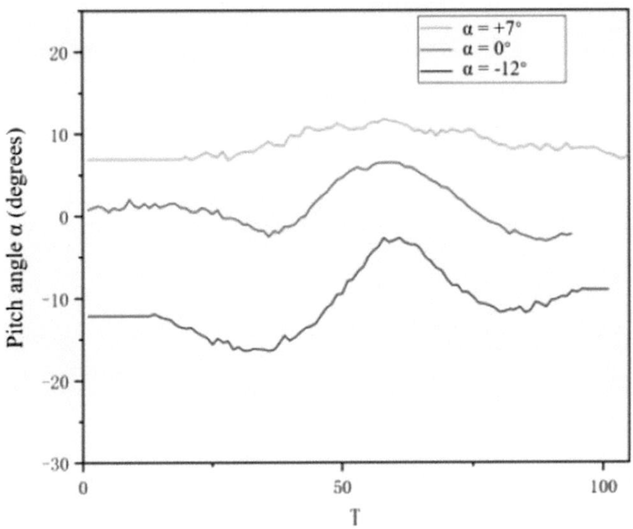

(b) Pitch angle variation

Fig. 8.31 Motion response of vehicles at different pitch angles encountering internal waves in the upper fluid layer

attitude is most unstable and more susceptible to the influence of internal solitary waves (Fig. 8.32).

(2) Interaction between self-propelled bodies with different pitch angles and internal solitary waves at the interface

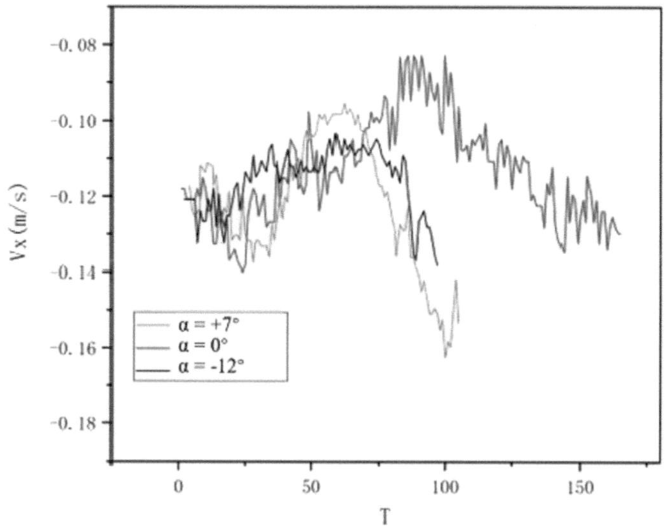

(a) Horizontal velocity

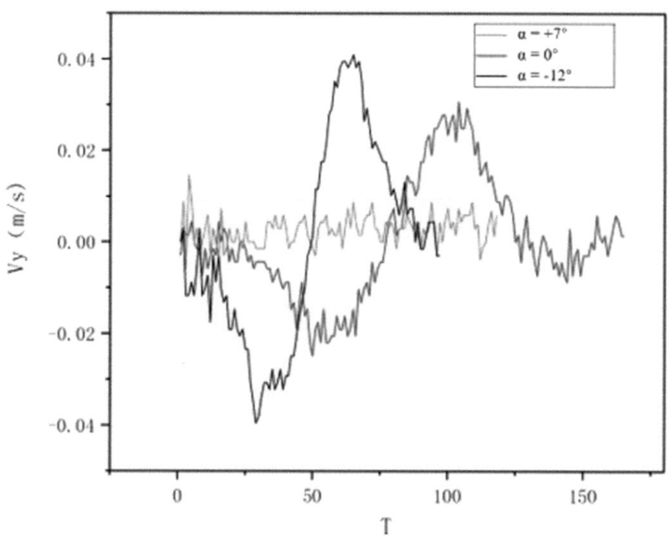

(b) Vertical velocity

Fig. 8.32 Velocity variation curves of self-propelled bodies with different pitch angles in upper layer fluid

This section focuses on Cases 20–22, studying self-propelled navigation bodies at the interface with three different navigation attitudes, as shown in Fig. 8.33. Maintaining consistent internal solitary wave amplitude, the study monitors changes in motion trajectory and state under wave action, obtaining through post-processing the centroid trajectory, pitch angle variation, and horizontal/vertical velocity curves of the self-propelled bodies.

Figure 8.34 displays the motion trajectories and pitch angle variations of self-propelled navigation bodies encountering internal waves at the density interface between upper and lower fluids under different motion attitudes. From Fig. 8.34a, it can be seen that all three initial attitude navigation bodies show basically consistent early-stage motion trajectories, all moving downward along the internal solitary wave surface. The vertical movement distance increases as the pitch angle changes from positive to zero to negative. After passing the wave trough, the upward-tilted and horizontal navigation bodies show basically consistent trajectory changes, moving upward with the wave surface. The navigation body with initial negative pitch angle doesn't follow the left wave surface upward but continues downward, experiencing significant depth loss, as shown in Fig. 8.35a.

Further analysis combined with Fig. 8.34b showing the variation curve of pitch angle reveals that: for the vehicle initially in horizontal position, its pitch angle follows the same trend as described above; for the vehicle initially in nose-up position, its pitch angle experiences a brief decrease when approaching the right wave surface, followed by a significant increase up to 20° (four times the initial angle), as shown in Fig. 8.35b. This occurs because the nose-up attitude places both ends of the vehicle in the upper fluid layer (with rightward surrounding velocity) while the tail extends into the lower layer (with leftward surrounding velocity), creating a clockwise torque on the vehicle. This also explains its shorter vertical travel distance and faster ascent. The pitch angle gradually recovers as the vehicle moves away from the internal solitary wave, with the final increase likely caused by wake effects. For the vehicle initially in nose-down position, its pitch angle shows minor reduction when passing the wave trough, then increases rapidly when approaching the left wave surface, reaching −28° (twice the initial angle), causing rapid depth loss. This demonstrates the high sensitivity of vehicle motion at the interface to pitch angle variations.

Post-processing of experimental results yielded the vehicle's horizontal and vertical velocity variation curves, shown in Fig. 8.36. Figure 8.36a reveals that the nose-up vehicle shows limited horizontal velocity increase, even experiencing deceleration near the wave trough. Analysis combining pitch angle variation shows that as pitch angle increases near the trough, most of the vehicle's nose remains in the upper fluid layer (with rightward flow velocity), while increased pitch angle enlarges the vehicle's projected area in vertical direction, enhancing flow resistance and causing deceleration. For vehicles initially in nose-up or nose-down positions, most areas remain in the lower fluid layer, so horizontal velocity continuously increases when approaching the internal wave, peaking at the trough. Figure 8.36b shows generally consistent vertical velocity trends with differences only in magnitude and period. Notably, the nose-down vehicle generates sustained downward velocity after passing

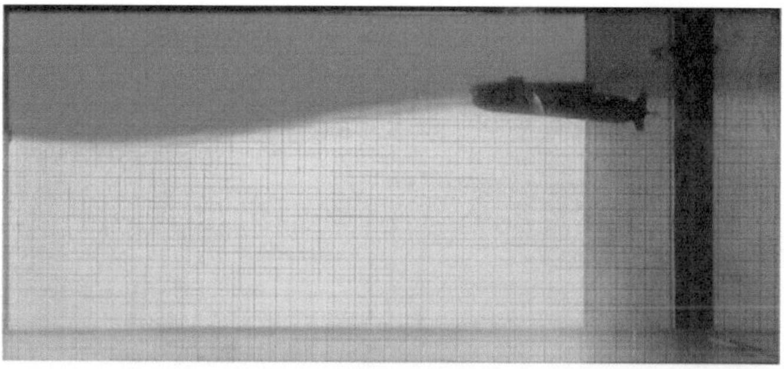

(a)Case 20

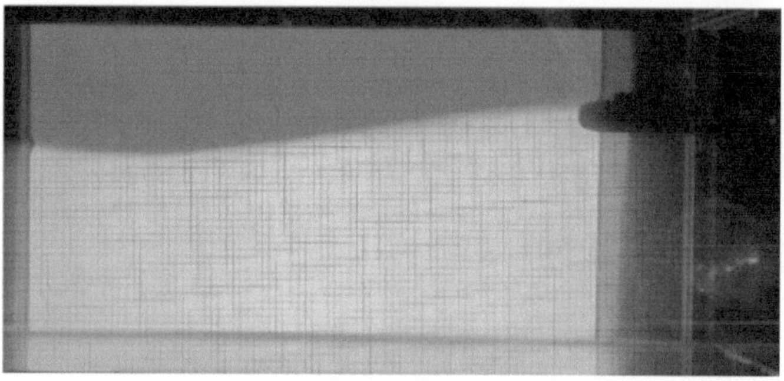

(b)Case 21

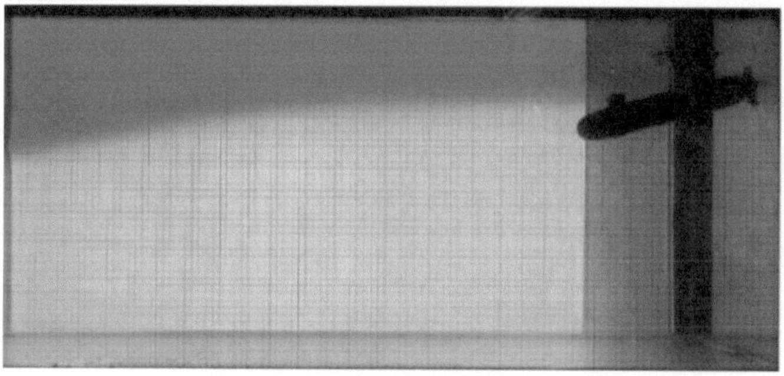

(c)Case 22

Fig. 8.33 Initial attitudes of navigation bodies with different pitch angles at the interface

the trough, further explaining the depth loss phenomenon in motion trajectory, which requires special caution.

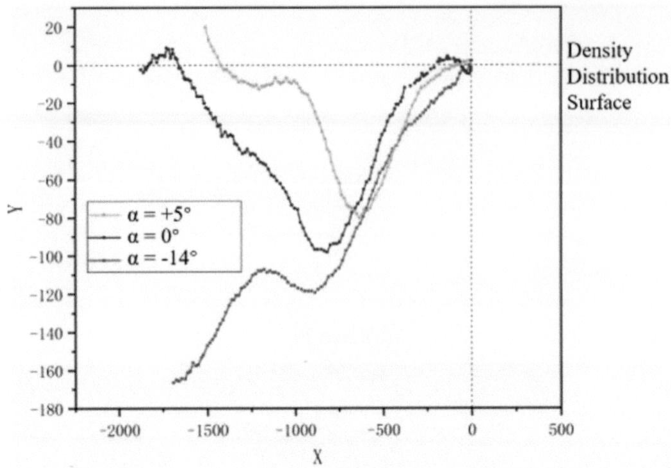

(a) Trajectory of center of mass

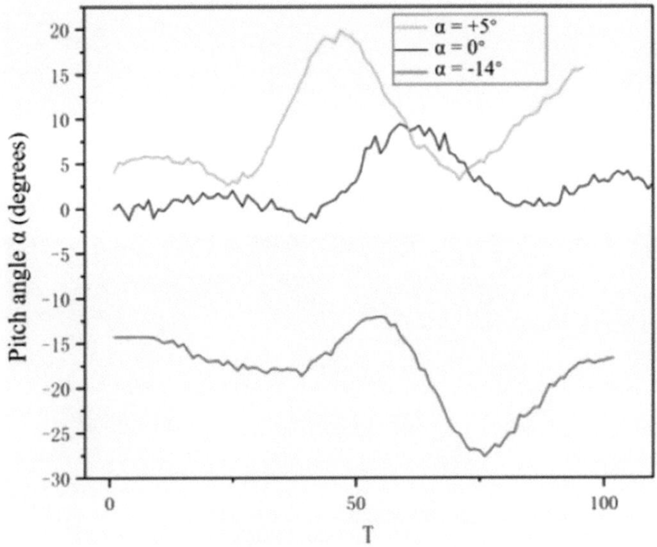

(b) Pitch angle variation

Fig. 8.34 Motion responses of vehicles with different initial pitch angles encountering internal waves at the interface

(a) Maximum pitch-down position

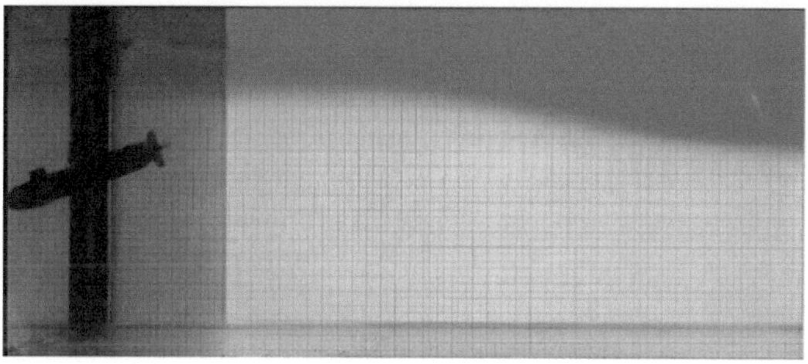

(b) Depth loss phenomenon

Fig. 8.35 Case 22 vehicle experiencing depth loss when moving away from internal solitary wave

8.5 Summary

Based on previous research conclusions regarding the interaction between internal solitary waves and the appendage performance of underwater vehicles, this chapter employs physical experiments to study the self-propulsion of underwater vehicles under internal solitary waves. A strongly stratified fluid environment was established in a stratified water tank, and large-amplitude internal solitary waves with controllable wave amplitudes were generated using the gravity collapse method. A series of experiments were conducted on the interaction between internal solitary waves and both suspended and self-propelled underwater vehicles. The motion response patterns of underwater vehicles under different conditions—such as submergence depth, wave amplitude, speed, and attitude—were obtained in an internal solitary wave environment. The main research findings include:

(1) A self-propelled underwater vehicle was designed and fabricated for small-scale internal solitary wave tanks, capable of hovering, depth control, speed control, and rudder angle control. Based on this, experimental, recording, and post-processing schemes were designed for studying the interaction between internal solitary waves and both suspended and self-propelled underwater vehicles.

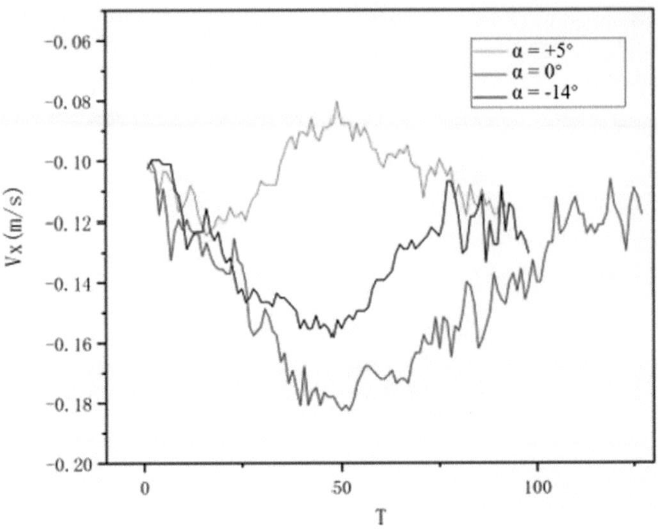

(a) Horizontal velocity

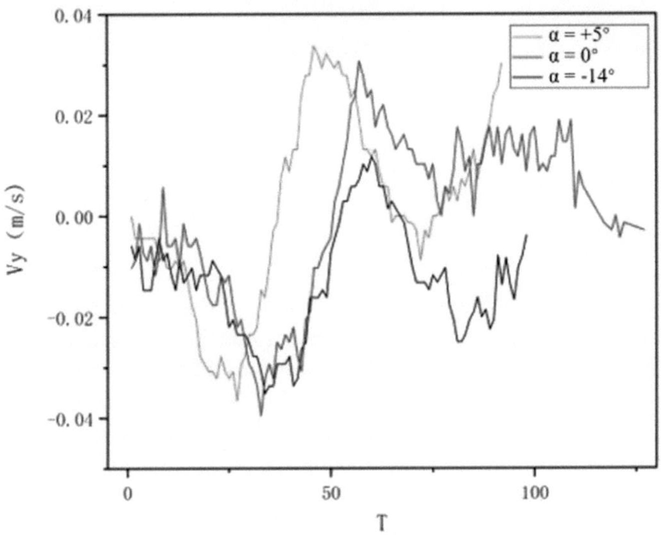

(b) Vertical velocity

Fig. 8.36 Velocity variation curves of self-propelled vehicles with different pitch angles at the interface

(2) For suspended underwater vehicles, the motion response varies significantly at different submergence depths. The primary influencing factor is the clockwise velocity field induced by internal solitary waves. When the vehicle is located below the trough of an internal solitary wave, it is more prone to experiencing "falling deep". For suspended vehicles at the interface, as the wave amplitude increases, the fluctuation in pitch angle also becomes larger, leading to greater instability in the vehicle's state, requiring more caution.
(3) For self-propelled underwater vehicles, wave amplitude also has a significant impact. When located in the upper fluid layer, as the wave amplitude increases, the depth loss of the vehicle decreases, but the variation in pitch angle becomes more drastic, with larger amplitude fluctuations, further exacerbating its attitude instability and posing a severe threat to safe navigation. However, when located at the density interface, the opposite conclusion holds: the larger the wave amplitude, the greater the range of motion in both horizontal and vertical directions. Yet, the attitude (pitch angle) of the self-propelled vehicle remains relatively stable at the interface, showing low sensitivity to changes in wave amplitude. The influence of speed cannot be ignored either. Experimental results indicate that low-speed vehicles are more prone to significant attitude changes under internal solitary waves, whereas high-speed vehicles can better resist hydrodynamic disturbances caused by internal solitary waves. Additionally, the pitch angle of the vehicle has a notable impact on its motion state in an internal solitary wave environment. When located in the upper fluid layer, vehicles with a nose-down attitude exhibit the largest variation range in vertical velocity and the most unstable motion, making them more susceptible to the effects of internal solitary waves. When positioned at the interface, the vehicle's motion state becomes more sensitive to changes in pitch angle. Vehicles initially in a nose-down state experience rapid depth loss after passing through the trough of an internal solitary wave, with the pitch angle during motion reaching twice the initial value.

Chapter 9
Concluding Remarks

The thesis, based on the engineering background of the interaction between oceanic internal solitary waves (ISWs) and underwater vehicles, conducts an in-depth study on the influence of ISWs on the motion response and load characteristics of underwater vehicles. It begins with an introduction to ISW theory and computational fluid dynamics (CFD) numerical methods, discussing the applicability of various theoretical equations. A three-dimensional numerical wave tank for strongly stratified and continuously stratified ISWs is established, analyzing the differences in flow fields between the two stratifications. The time-step independence of wave generation is further validated, and the structured grid of the vehicle is verified for grid independence with local optimization. The accuracy of the ISW load and fluid–structure coupling model is demonstrated through a case study of ISW interaction with a suspended sphere.

Building on this, simulations of interactions between strongly stratified ISWs and suspended/moving vehicles under different conditions are conducted, investigating the effects of submergence depth, wave amplitude, and speed on the vehicle's motion response and load characteristics. Similar simulations are performed for continuously stratified ISWs, and the differences between the two stratifications are compared. Numerical simulations are also carried out to study the interaction between stratified structures/ISWs and propellers/rudders under various conditions, analyzing the influence of parameters such as wave amplitude, submergence depth, and rudder angle on the hydrodynamic characteristics of appendages. Finally, a series of experiments involving ISWs interacting with suspended and self-propelled vehicles are conducted, studying the influence of submergence depth, wave amplitude, speed, and pitch angle on the motion response and state of the vehicle in an ISW environment. These findings provide a reference for the control strategies of underwater vehicles encountering ISWs in real-world conditions.

The main research results are as follows:

(1) The theories and applicability of three types of ISWs—KdV, MCC, and DJL—are discussed, comparing their waveforms, wave speeds, and characteristic wavelengths

P. Du et al., *Internal Wave Hydrodynamics on Submersibles*,
https://doi.org/10.1007/978-981-95-4479-0_9

under current fluid parameters. The results show that the MCC waveform aligns well with the density median waveform of the DJL equation, with its characteristic wavelength lying between those of the KdV and mKdV theoretical solutions. The theoretical wave speed of the DJL equation is lower than that of the strongly stratified case.

(2) Based on the RANS equations and the standard k-ε turbulence model, a three-dimensional numerical wave tank for strongly and continuously stratified ISWs is established. The horizontal velocity fields of the two stratifications are compared, and the time-step independence of wave generation is analyzed. The results indicate that, unlike strongly stratified ISWs, the horizontal velocity of continuously stratified ISWs undergoes a gradual change at the density interface, with smooth and symmetrical streamlines and no trailing waves. Additionally, three sets of structured grids are generated for a full-scale vehicle model, and grid independence is verified through surface pressure and friction coefficients, confirming the reliability of overset grids. The grids near the sail and rudder are optimized based on y + values. The accuracy of the ISW load and fluid–structure coupling model is validated through a case study of ISW interaction with a suspended sphere, laying the foundation for subsequent simulations of ISW-vehicle coupling.

(3) Simulations of interactions between strongly stratified ISWs and suspended/moving vehicles under different conditions are conducted, studying the effects of submergence depth, wave amplitude, and speed on the vehicle's motion response and load characteristics. The results show that vehicles located above or at the density interface experience significant motion responses, with closer proximity to the interface leading to greater load effects. Wave amplitude primarily affects the speed of motion response and peak load magnitude—larger amplitudes result in faster responses and higher peak loads. Moving vehicles are more prone to loss of control under ISW influence, with sharp increases in forces. Higher speeds lead to larger peak loads but shorter interaction durations.

(4) Simulations of interactions between continuously stratified ISWs and suspended/moving vehicles under different conditions are performed to study the influence of submergence depth, wave amplitude, and speed on the vehicle's motion response and load characteristics in real oceanic environments. The differences between strongly and continuously stratified ISWs are summarized. The results indicate that the influence trends of submergence depth, wave amplitude, and speed are similar to those in strongly stratified cases. However, under continuously stratified ISWs, the vehicle's motion and forces exhibit stronger oscillations, slower response rates, and lower peak loads compared to strongly stratified cases.

(5) Using the established ISW numerical wave tank, the effects of continuously stratified structures and ISWs on propeller hydrodynamic performance are studied. The results show that density stratification has a minor impact on propeller performance. For advance coefficients $J \leq 1.0$, the thrust coefficient, torque coefficient, and open-water efficiency closely match those in uniform flow. Only when $J \geq 1.0$ do

these parameters slightly exceed those in non-stratified cases, with the most significant impact on open-water efficiency. In contrast, ISWs significantly affect propeller performance and wake fields, particularly when the propeller traverses the front wave face or is above the wave trough, where thrust and torque drop sharply (up to 22.42% compared to initial thrust), posing a major threat to navigation safety.

(6) Numerical simulations of ISW-rudder interactions are conducted, with open-water rudder performance simulations validating the accuracy and reliability of the method. The stall angle of the rudder is found to be around 35°. The effects of submergence depth, rudder angle, and wave amplitude on rudder hydrodynamic performance in ISW environments are studied. The results show that rudders located below the wave surface experience severe force oscillations and amplitudes far exceeding those at other depths, along with frequent flow separation, posing significant risks. Rudder angle has the most pronounced effect—under ISWs, stall angles occur much earlier at all depths, suggesting that rudder angles should be kept below 15° (ideally around 10°) when encountering ISWs. Wave amplitude has minimal influence on the force variation pattern.

(7) Physical experiments are conducted to study self-propelled vehicle motion under ISWs. A strongly stratified fluid environment is created in a stratified flow tank, and ISWs with controllable amplitudes are generated using the gravity collapse method. A small-scale vehicle suitable for ISW environments is designed and tested to investigate the motion response under varying submergence depths, wave amplitudes, speeds, and attitudes. The findings reveal that for suspended vehicles, "depth plunging" is more likely when located below the ISW trough. For self-propelled vehicles in the upper layer, larger wave amplitudes reduce depth plunging but increase pitch angle fluctuations, with vehicles in a bow-down attitude exhibiting the largest vertical velocity variations and the most unstable motion. At the density interface, larger wave amplitudes expand motion ranges in both horizontal and vertical directions, with motion states being more sensitive to pitch angle changes. Bow-down vehicles experience rapid depth plunging after passing the trough, with pitch angles doubling the initial value. Additionally, low-speed vehicles undergo larger attitude changes under ISWs, while high-speed vehicles better resist ISW-induced hydrodynamic disturbances.

References

1. Alford, M, Peacock, T, MacKinnon J, et al. The formation and fate of internal waves in the South China Sea[J]. Nature 521, 2015: 65–69.
2. Alin N, Bensow R E, Fureby C, et al. Current Capabilities of DES and LES for Submarines at Straight Course[J]. Journal of Ship Research, 2010, 54(3):184-196.
3. Apel J R, Byrne H M, Proni J R, et al. Observations of oceanic internal and surface waves from the Earth Resources Technology Satellite[J]. Journal of Geophysical Research, 1975, 80(6): 865-881.
4. Bao S, Meng J, Sun L, et al. Detection of ocean internal waves based on Faster R-CNN in SAR images[J]. Journal of Oceanology and Limnology, 2020, 38(1):55-63.
5. Benjamin, Brooke T. Internal waves of permanent form in fluids of great depth[J]. Journal of Fluid Mechanics, 1967, 29(3): 559–592.
6. Benjamin T B. Internal waves of finite amplitude and permanent form[J]. Journal of Fluid Mechanics, 1966, 25(2): 241-270.
7. Benney D J. Long Non-Linear Waves in Fluid Flows[J]. Studies in Applied Mathematics, 1966, 45(1-4):52–63.
8. Cai S Q, Gan Z J. Research progress on solitary internal waves in the northern South China Sea [J]. Advances in Earth Science, 2001, 16(2):215-219.
9. Cai S Q, Long X M, Gan Z J. Preliminary Study on the Forces of Soliton Internal Waves on Small-Diameter Cylindrical Piles [J]. Advances in Hydrodynamics (Series A), 2002(04):497-506.
10. Cai S, Long X, Gan Z. A method to estimate the forces exerted by internal solitons on cylindrical piles[J]. Ocean Engineering, 2003, 30(5): 673-689.
11. Cai S, Wang S, Long X. A simple estimation of the force exerted by internal solitons on cylindrical piles[J]. Ocean Engineering, 2006, 33(7): 974-980.
12. Cai S, Long X, Wang S. Forces and torques exerted by internal solitons in shear flows on cylindrical piles[J]. Applied Ocean Research, 2008, 30(1): 72-77.
13. Cai, S, Xie J, He J. An Overview of Internal Solitary Waves in the South China Sea[J]. Surv Geophys. 2012, 33, 927-943.
14. Camassa R, Choi W, Michallet H, et al. On the realm of validity of strongly nonlinear asymptotic approximations for internal waves[J]. Journal of Fluid Mechanics, 2006, 549: 1-23.
15. Carr M, King S E, Dritschel D G. Numerical simulation of shear-induced instabilities in internal solitary waves[J]. Journal of Fluid Mechanics, 2011, 683: 263-288.
16. Chen C Y. An experimental study of stratified mixing caused by internal solitary waves in a two-layered flfluid system over variable seabed topography[J]. Ocean Engineering. 2007, 34:1995-2008.

P. Du et al., *Internal Wave Hydrodynamics on Submersibles*,
https://doi.org/10.1007/978-981-95-4479-0

17. Chen C Y, Hsu J R C, Chen C W, et al. Generation of internal solitary wave by gravity collapse[J]. Journal of Marine Science and Technology, 2007, 15(1): 1.
18. Chen J, You Y X, Liu X D, et al. Numerical Simulation of the Interaction between Internal Solitary Waves and Submerged Bodies with Velocity [J]. Advances in Hydrodynamics (Series A), 2010, 25(3): 344-351.
19. Chen M, Chen K, You Y X. Experimental investigation of internal solitary wave forces on a semi-submersible[J]. Ocean Engineering, 2017, 141: 205-214.
20. Chen M, Chen K, You Y X, et al. Experimental study of forces on a multi-column floating platform in internal solitary waves[J]. Applied Ocean Research, 2018, 78:192-200.
21. Cheng M H, Hwang R R, Hsieh C M. Numerical study on the transformation of an internal solitary wave propagating across a vertical cylinder[J]. Applied Ocean Research, 2020, 95: 102016.
22. Choi W, Camassa R. Fully nonlinear internal waves in a two-fluid system[J]. Journal of Fluid Mechanics. 1999,396: 1-36.
23. Cui J, Dong S, Wang Z, et al. Kinematic response of submerged structures under the action of internal solitary waves[J]. Ocean Engineering, 2020, 196: 106814.
24. Cui J, Dong S, Wang Z. Study on applicability of internal solitary wave theories by theoretical and numerical method[J]. Applied Ocean Research, 2021, 111: 102629.
25. Cui J, Sheng D, Wang Z F, Cheng J Y, Ren J Y. Experimental study on motion response of a submersible spherical model under the action of internal solitary wave propagating over a ridge topography[J]. Ocean Engineering, 2022, 258, 111700.
26. Dauxois T, Peyrard M. Physics of solitons[M]. Cambridge University Press, 2006, chap.1.
27. Deepwell D, Clarry C, Subich C, et al. Vortex generation due to internal solitary wave propagation past a sidewall constriction[J]. Journal of Fluid Mechanics, 2021, 913.
28. Ding W Y. Numerical Simulation Study on the Interaction between Internal Solitary Waves and Marine Structures [D]. Dalian University of Technology, 2020.
29. Drazin P G, Johnson R S. Solitons: an introduction[M]. Cambridge university press, 1989.
30. Du H, Wei G, Gu M, et al. Experimental investigation of the load exerted by nonstationary internal solitary waves on a submerged slender body over a slope[J]. Applied Ocean Research, 2016: 216–223.
31. Du H, Wei G, Zeng W H, et al. Experimental Study on the Influence of Downward Convex Internal Solitary Waves on the Motion Characteristics of Slender Submerged Bodies [J]. Journal of Ship Mechanics, 2017, 21(10):8.
32. Du H, Wei G, Wang S D, et al. Experimental study of elevation-and depression-type internal solitary waves generated by gravity collapse[J]. Physics of Fluids, 2019, 31(10): 102104.
33. Duda T F, Lynch J F, Irish J D, et al. Internal tide and nonlinear internal wave behavior at the continental slope in the northern south China Sea[J]. IEEE, 2004(4): 1105-1129.
34. Dunphy M, Subich C, Stastna M. Spectral methods for internal waves: indistinguishable density profiles and double-humped solitary waves[J]. Nonlinear Processes in Geophysics, 2011, 18(3): 351-358.
35. Fang X H, Du T. Ocean Internal Waves and Internal Waves in China Sea [M]. Ocean University of China Press, 2005.
36. Ferziger J H, Perić M, Street R L. Computational methods for fluid dynamics[M]. Berlin: springer, 2002.
37. Fructus D, Grue J. Fully nonlinear solitary waves in a layered stratified fluid[J]. Journal of Fluid Mechanics, 2004, 505:323-347.
38. Fu L L. Seasat views oceans and sea ice with synthetic-aperture radar[M]. California Institute of Technology, Jet Propulsion Laboratory, 1982.
39. Funakoshi M, Oikawam. Long internal waves of large amplitude in a two-layer fluid.Journal of the Physical Society of Japan[J], 1986,55:128–144.
40. Gong Y K, Xie J S, Xu J X, Chen Z W, He Y H, Cai S Q. Oceanic internal solitary waves at the Indonesian submarine wreckage site[J]. Acta Oceanologica Sinica, 2022, 41(3): 109-113.
41. Grimshaw R, Guo C, Helfrich K, et al. Combined effect of rotation and topography on shoaling oceanic internal solitary waves[J]. Journal of Physical Oceanography, 2014, 44(4): 1116-1132.

42. Grimshaw R, Wang C X, Li L. Modelling of polarity change in a nonlinear internal wave train in Laoshan Bay[J]. Journal of Physical Oceanography, 2016, 46(3): 965-974
43. Gu M M, Wei G, Deng B, Du H, Wu J L. Evolution of Internal Solitary Waves Over a Ridge and Their Force Characteristics on Slender Submerged Bodies [J]. Acta Mechanica Sinica, 2017, 49(6): 1260-1271.
44. Guan H, Wei G, Du H. Hydrodynamic Characteristics of the Interaction between Internal Solitary Waves and Submarines [J]. Journal of PLA University of Science and Technology (Natural Science Edition), 2012, 13(5): 577-582.
45. Hartharn-Evans S G, Carr M, Stastna M, et al. Stratification effects on shoaling internal solitary waves[J]. Journal of Fluid Mechanics, 2022, 933.
46. Helfrich K R, Melville W K. Long nonlinear internal waves. Annual Review of Fluid Mechanics[J]. 2006, 38:395–425.
47. Horn D A, Imberger J, et al. The degeneration of large-scale interfacial gravity waves in lakes[J]. Journal of Fluid Mechanics, 2001, 434, 181–207.
48. Hu B, Pan G, Du X X, Huang Q G, Wang Y Y. Numerical Study on Submarine Viscous Flow Field Based on Block Grid Technology [J]. Journal of Northwestern Polytechnical University, 2012, 30(05): 689-693.
49. Hu Y, Zou L, Ma X, et al. Numerical Simulation for the Evolution of Internal Solitary Waves Propagating over Slope Topography[J]. Journal of Marine Science and Engineering, 2021, 9(11): 1224.
50. Huang W H, You Y X, Wang X, et al. Experimental Study on Internal Solitary Wave Loads and Theoretical Models for Cylindrical Structures [J]. Acta Mechanica Sinica, 2013, 45(5): 716-728.
51. Huang X, Zhao W, Tian J, et al. Mooring observations of internal solitary waves in the deep basin west of Luzon Strait[J]. Acta Oceanologica Sinica, 2014, 33: 82-89.
52. Huang X, Chen Z, Zhao W, Zhang Z, Zhou C, Yang Q, Tian J. An extreme internal solitary wave event observed in the northern South China Sea[J]. Scientific Reports. 2016, 6, 30041.
53. Huang M M, Zhang N, Zhu A J. Study on Hydrodynamic Loads and Motion Characteristics of Underwater Vehicles Under the Action of Internal Waves [J]. Journal of Ship Mechanics, 2019, 23(5): 531-540.
54. Huang X, Huang S W, Zhao W, Zhang Z W, Zhou C, Tian J W. Temporal variability of internal solitary waves in the northern South China Sea revealed by long-term mooring observations[J]. Progress in Oceanography, 2022, 201, 102716.
55. Hui Du, Gang Wei, Wang S D, Wang X L. Experimental study of elevationand depression-type internal solitary waves generated by gravity collapse[J]. Physics of Fluids. 2019, 31, 102104.
56. Iwata R, Nakayama K, Sato T, et al. Effect of Pycnocline Thickness on Internal Solitary Wave Breaking Over a Slope and its Classification[J]. Journal of Japan Society of Civil Engineers, Ser. B1 (Hydraulic Engineering), 2021, 77(1): 59–73.
57. Jackson C. Internal wave detection using the Moderate Resolution Imaging Spectroradiometer (MODIS)[J]. Jounal of Geophysical Research, 2007, 112:1-13.
58. Jackson C R, Apel J. An atlas of internal solitary-like waves and their properties[J]. Contract, 2004, 14(03-C): 0176.
59. Ji M, Chen K, You Y, et al. Experimental study of wave loading by internal solitary waves on a submerged slender body[C]//International Conference on Offshore Mechanics and Arctic Engineering. American Society of Mechanical Engineers, 2020, 84386: V06BT06A059.
60. John D. Anderson. Computational Fluid Dynamics: The Basics with Applications [M]. Translated by Wu S P, Liu Z M. Beijing: Mechanical Industry Press, 2007: 26–51.
61. Klymak, J M, Pinkel, R, Liu C T, Liu A K, David L. Prototypical solitons in the South China Sea[J]. Geophys. Res. Lett. 2006, 33, L11607.
62. Lamb K G. Numerical experiments of internal wave generation by strong tidal flow across a finite amplitude bank edge[J]. Journal of Geophysical Research: Oceans, 1994, 99(C1): 843-864.
63. Lamb K G. A numerical investigation of solitary internal waves with trapped cores formed via shoaling[J]. Journal of Fluid Mechanics. 2002, 451: 109-144.

64. Lamb K G. Energy and pseudoenergy flux in the internal wave field generated by tidal flow over topography[J]. Continental shelf research, 2007, 27(9): 1208-1232.
65. Lamb K G, Warn-Varnas A. Two-dimensional numerical simulations of shoaling internal solitary waves at the ASIAEX site in the South China Sea[J]. Nonlinear Processes in Geophysics, 2015, 22(3): 289-312.
66. Li J C. Underwater Waves—Ocean Internal Waves [J]. Mechanics in Engineering, 2005, 27(2):1-6.
67. Li J Y, Zhang Q H, Chen T Q. Numerical Simulation of Internal Solitary Waves in Continuously Stratified Fluids [J]. Journal of Tianjin University: Science and Engineering, 54(2):10.
68. Li J, Zhang Q, Chen T. Numerical Investigation of Internal Solitary Wave Forces on Submarines in Continuously Stratified Fluids[J]. Journal of Marine Science and Engineering. 2021, 9, 1374.
69. Li J, Zhang Q, Chen T. Numerical Investigation of Internal Solitary Wave Forces on Submarines in Continuously Stratified Fluids[J]. Journal of Marine Science and Engineering, 2021, 9(12): 1374.
70. Li J, Zhang Q, Chen T. ISWFoam: A numerical model for internal solitary wave simulation in continuously stratified fluids[J]. Geoscientific Model Development, 2022, 15(1): 105-127.
71. Lien R C, Tang T Y, Chang M H, et al. Energy of nonlinear internal waves in the South China Sea[J]. Geophysical Research Letters, 2005, 32(5).
72. Lien R C, Henyey F, Ma B, et al. Large-amplitude internal solitary waves observed in the Northern South China Sea: properties and energetics[J]. Journal of Physical Oceanography, 2014, 44(4): 1095-1115.
73. Liu H L, Huang T T. Summary of DARPA SUBOFF experimental program data[R]. Naval Surface Warfare Center Carderock Div Bethesda MD Hydromechanics Directorate, 1998.
74. Manual U D F. ANSYS FLUENT 12.0[J]. Theory Guide, 2009.
75. Meyers R. Mathematics of complexity and dynamical systems[M]. Springer Science & Business Media, 2011, 1561–1574.
76. Michallet H, Barthelemy E. Experimental study of interfacial solitary waves[J]. Journal of Fluid Mechanics. 1998, 366:159-177.
77. Morison J R, Johnson J W, Schaaf S A. The force exerted by surface waves on piles[J]. Journal of Petroleum Technology, 1950, 2(05): 149-154.
78. Norman F, Polmar N C. “submarine"[OL]. Encyclopedia Britannica, https://www.britannica.com/technology/submarine-naval-vessel, 2023.
79. Ono H. Wave propagation in an inhomogeneous anharmonic lattice[J]. Journal of the Physical Society of Japan, 1972, 32(2): 332-336.
80. Orr M H, Mignerey P C. Nonlinear internal waves in the South China Sea: Observation of the conversion of depression internal waves to elevation internal waves[J]. J Geophys Res, 2003, 108:3064.
81. Osborne A R, Burch T L. Internal solitons in the Andaman Sea[J]. Science, 1980, 208(4443): 451-460.
82. Qiu L Y. Numerical Simulation of Submarine Viscous Flow Field and Methods for Drag Prediction [D]. Huazhong University of Science and Technology, 2006.
83. Ramp S R, Tang T Y, Duda T F, et al. Internal solitons in the northeastern South China Sea. Part I: Sources and deep water propagation[J]. IEEE Journal of Oceanic Engineering, 2004, 29(4): 1157–1181.
84. Shi S D. Submarine Maneuverability [M]. National Defense Industry Press, 1995.
85. Simmons H, Chang M H, Chang Y T, et al. Modeling and prediction of internal waves inthe South China Sea[J]. Oceanography, 2011, 24(4):88-99.
86. Song Z J, Teng B, Gou Y, et al. Comparisons of internal solitary wave and surface wave actions on marine structures and their responses[J]. Applied Ocean Research, 2011, 33(2): 120-129.
87. Stastna M. Internal Waves in the Ocean[M]. Springer Nature Switzerland AG. 2022.
88. Stastna M, Lamb K G. Large fully nonlinear internal solitary waves: The effect of background current[J]. Physics of Fluids. 2002: 2987–2999.

89. Tian X. A Summary of the Korteweg-de Vries Equation-tianxiang[M]. 2015.
90. Tian Z, Chen T, Yu L, et al. Penetration depth of the dynamic response of seabed induced by internal solitary waves[J]. Applied Ocean Research, 2019, 90: 101867.
91. Vlasenko V, Stashchuk N. Three-dimensional shoaling of large-amplitude internal waves[J]. Journal of Geophysical Research: Oceans, 2007, 112.
92. Vlasenko V, Stashchuk N, Guo C, Chen X. Multimodal structure of baroclinic tides in the South China Sea[J]. Nonlinear Processes in Geophysics, 2010, 17(5): 529-543
93. Walker L R. Interfacial solitary waves in a two-fluid medium[J]. Physics of Fluids, 1973, 16(11):1796-1804.
94. Wang R Y. Numerical study on the effect of internal solitary waves on submersible structures [D]. Dalian University of Technology, 2018.
95. Wang Z. Research on the Characteristics of Ocean Solitary Waves in No-Flow and Flow Conditions [D]. Harbin Engineering University, 2021.
96. Wang X, Lin Z Y, You Y X. Numerical Simulation of Internal Solitary Wave Load Characteristics on Semi-Submerged Platforms [J]. Journal of Ship Mechanics, 2015, 19(IO):1173–1185.
97. Wang L L, Wang Y, Wei G, et al. Characteristics of Forces on Cylinders and Square Columns in Internal Solitary Wave Environment—Part I: Physical Experiments [J]. Advances in Water Science, 2017, 28(3): 429-437.
98. Wang L L, Wang Y, Wei G, et al. Characteristics of Forces on Cylinders and Square Columns in Internal Solitary Wave Environment—Part II: Numerical Simulations [J]. Advances in Water Science, 2017, 28(04): 112-121.
99. Wang S, Wei G, Du H, et al. Experimental investigation of the wave-flow structure of an oblique internal solitary wave and its force exerted on a slender body[J]. Ocean Engineering, 2020, 201: 107057.
100. Wang Z, Wang Z, Yuan C. Oceanic internal solitary waves in three-layer fluids of great depth[J]. Acta Mechanica Sinica, 2022, 38(2): 1-14.
101. Wang C, Huang B, Xu T, et al. Numerical modeling of energy dissipation of internal solitary waves encountering step topography[J]. Ocean Engineering, 2022, 259: 111853.
102. Wei G, Du H, Xu X H, et al. Experimental investigation of the generation of large-amplitude internal solitary wave and its interaction with a submerged slender body[J]. Science China Physics, Mechanics and Astronomy, 2014, 57(2): 301-310.
103. Wei G, Du H, Xu X, et al. Experimental investigation of the generation of large-amplitude internal solitary wave and its interaction with a submerged slender body[J]. Science China-physics Mechanics & Astronomy, 2014, 57(2): 301-310.
104. Wu J L, Wei G, Du H. Experimental Study on the Interaction Characteristics of Downward Convex Internal Solitary Wave Flow Field and Horizontally Placed Slender Submerged Bodies [J]. Advances in Hydrodynamics (Series A), 2017, 32(5): 592-599.
105. Xie J, Xu J, Cai S. A numerical study of the load on cylindrical piles exerted by internal solitary waves[J]. Journal of Fluids and Structures, 2011, 27(8): 1252-1261.
106. Xu Z Y. Dynamics of Ocean Internal Waves [M]. Science Press, 1999.
107. Xu Z H, Liu K, Yin B S, et al. Long-range propagation and associated variability of internal tides in the South China Sea[J]. Journal of Geophysical Research: Oceans, 2016, 121(11): 8268-8286.
108. Xu J, He Y, Chen Z, et al. Observations of different effects of an anti-cyclonic eddy on internal solitary waves in the South China Sea[J]. Progress in Oceanography, 2020, 188: 102422.
109. Yang F. Study on the Interaction between Ocean Internal Solitary Waves and Underwater Vehicles [D]. Jiangsu University of Science and Technology.
110. Yang Y J, Tang T Y, Chang M H, et al. Solitons northeast of Tung-Sha Island during the ASIAEX pilot studies[J]. IEEE Journal of Oceanic Engineering, 2004, 29(4): 1182-1199.
111. Yousefi Y, Muminov K K. A simple classification of solitons[J]. arXiv:1206.1294, 2012.
112. Yu D, Zhang Z, Dong H, et al. A novel dynamic model and the oblique interaction for ocean internal solitary waves[J]. Nonlinear Dynamics, 2022, 108(1): 491-504.
113. Yuan Y T. Characteristics of Flow Fields Induced by Interfacial Waves in Two-Layer Fluid[J]. New trends in fluid mechanics research. 2007, 09:368-371.

114. Zeng Y, Wang H B, Sun M B, et al. Review of Improvements in SST Turbulence Models [J]. Acta Aeronautica et Astronautica Sinica, 2023, 44: 027411.
115. Zhang L. Dynamic Characteristics and Dynamic Response of Internal Solitary Waves on Deep-Sea Risers [D]. Ocean University of China, 2013.
116. Zhang R, Chen K, You Y. Experimental, numerical and simplified theoretical model study for internal solitary wave load on FPSO with emphasis on scale effect[J]. China Ocean Engineering, 2019, 33(1): 26-33.
117. Ziegenbein J. Short internal waves in the Strait of Gibraltar[C]//Deep Sea Research and Oceanographic Abstracts. Elsevier, 1969, 16(5): 479-487.
118. Zou L, Du B, Ma X Y, et al. Study on the applicability of internal solitary wave theory models [J]. Proceedings of the Chinese Congress of Theoretical and Applied Mechanics (CCTAM 2019). Hangzhou, 2019, 13.

MIX
Papier aus verantwortungsvollen Quellen
Paper from responsible sources
FSC® C105338

If you have any concerns about our products,
you can contact us on
ProductSafety@springernature.com

In case Publisher is established outside the EU,
the EU authorized representative is:
Springer Nature Customer Service Center GmbH
Europaplatz 3, 69115 Heidelberg, Germany

Printed by Libri Plureos GmbH
in Hamburg, Germany